W9-ATC-528

2021 NIGHT SKY ALMANAC

A Month-by-Month Guide to North
America's Skies from The Royal
Astronomical Society of Canada

Nicole Mortillaro

FIREFLY BOOKS

A Firefly Book

Published by Firefly Books Ltd. 2020
Copyright © 2020 Firefly Books Ltd.
Copyright © 2020 The Royal Astronomical Society of Canada
Text © 2020 Nicole Mortillaro
Photographs © as listed on page 119

All rights reserved. No part of this publication may be reproduced, stored in a retrieval system, or transmitted in any form or by any means, electronic, mechanical, photocopying, recording or otherwise, without the prior written permission of the Publisher.

Second printing, 2021

Library of Congress Control Number: 2020941882

Library and Archives Canada Cataloguing in Publication
Title: 2021 night sky almanac : a month-by-month guide to North America's skies
    from the Royal Astronomical Society of Canada / Nicole Mortillaro.
Other titles: Two thousand twenty-one night sky almanac |
    Twenty twenty-one night sky almanac
Names: Mortillaro, Nicole, 1972- author. |
    Royal Astronomical Society of Canada, issuing body.
Description: Includes bibliographical references.
Identifiers: Canadiana 20200288091 | ISBN 9780228102595 (softcover)
Subjects: LCSH: Astronomy—Canada—Observers' manuals. |
    LCSH: Astronomy—Canada—Amateurs' manuals. |
    LCSH: Astronomy—United States—Observers' manuals. |
    LCSH: Astronomy—United States—Amateurs' manuals. |
    LCSH: Astronomy—Popular works. | LCGFT: Handbooks and manuals.
Classification: LCC QB64 .M67 2020 | DDC 523—dc23

Published in Canada by
Firefly Books Ltd.
50 Staples Avenue, Unit 1
Richmond Hill, Ontario L4B 0A7

Published in the United States by
Firefly Books (U.S.) Inc.
P.O. Box 1338, Ellicott Station
Buffalo, New York 14205

Project manager/Editor (Firefly Books): Julie Takasaki
Project manager (RASC): Robyn Foret, President of The RASC
Copyeditor: Elizabeth Howell
Technical editors: David M.F. Chapman, Fellow of The RASC,
    and James S. Edgar, Editor, RASC *Observer's Handbook*
Graphic design: Hartley Millson, Noor Majeed and Stacey Cho
Illustrations and sky charts: Peter Kovalik
Printed in Canada

 We acknowledge the financial support of the Government of Canada.

Contents

Introduction

If you're reading this, you clearly love the night sky and all the joy it brings you.

This guide aims to provide novice and intermediate amateur astronomers with the knowledge they need to enjoy the night sky in 2021, from understanding planets, nebulae, comets, asteroids and meteors, to learning about annual and significant celestial events. It's your pocket guide to the cosmos.

But first, it's important that you understand how we navigate the night sky. Just like for navigation on Earth, astronomers use particular coordinates in the sky to figure out what we're looking at.

The **celestial sphere** is an imaginary sphere with Earth at its center. At any one time, an observer's night-sky view only includes half of this sphere, because the other half is below the horizon.

Earth's axis is tilted at 23.5 degrees to the **plane** of the Solar System — the plane being the orbit of the Earth around the Sun. For observers on the ground, the celestial sphere seems to rotate from east to west; that is why the Sun, for example, rises in the east and sets in the west.

The Earth's axis points less than 1 degree away from **Polaris**, the North Star, in the Northern Hemisphere. It's really important to know where north is, so you can navigate around the sky. Use a star chart, your smartphone or both to find true north at your observing location. (In the south, the Earth's axis points about one

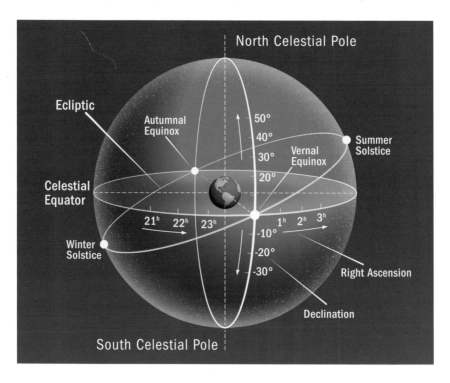

degree away from **Sigma Octantis**, a faint star; given Canada and the United States are in the Northern Hemisphere, however, we will focus on Polaris). When you look at the sky over long periods of time, Polaris stays still as the other stars rotate around it.

You can easily see this phenomenon by setting up a camera, pointing it toward Polaris, and taking an extended (or long-exposure) photograph of the stars. You will see circular streaks in the sky as other stars rotate around Polaris. Some stars are so close to the North Celestial Pole that they never set. Other stars farther from the pole will rise in the east and set in the west, just like our Sun. Many stars are so far from the pole that they never rise above your horizon.

While beginner astronomers will navigate using a simplified star chart, it's helpful to know some of the terminology we use for finding our way around the sky. That way, as you gain confidence, you can follow professional astronomers in their observations.

The point directly overhead your observing location is called the **zenith**, and the **celestial equator** is directly above the Earth's equator. The point directly below you (under the horizon and opposite to the zenith) is the **nadir**. The line that runs from the north point on the horizon, through the zenith, to the south point on the horizon, is the **meridian**.

To navigate the night sky, astronomers created a type of latitude and longitude called **celestial coordinates**. In astronomy, we use **declination** (Dec.) and **right ascension** (RA). Declination is like latitude on Earth, running from north to south. Right ascension is like longitude on Earth, running from east to west. Declination is measured in degrees, and right ascension is measured in hours, minutes and seconds.

Star trails with Polaris at the center

As Earth rotates, the apparent position of most of the stars changes. More advanced astronomers may want to learn more about exactly where to find faint stars in their telescopes. For these situations, we navigate using **altitude** (angular elevation above the horizon, between 0 degrees at the horizon and 90 degrees at the zenith) and **azimuth** (the number of degrees clockwise from due north). But if you're just starting out, don't worry yet about these advanced observing techniques. A simple star chart will help you find the constellations and the planets.

If you want to look for planets, you should also pay attention to the **ecliptic**. That is the path the Sun takes through the constellations, and the Moon and planets do not stray far from that path.

Handy Sky Measures

Astronomers need to know how far apart things are in the sky. And they do this using angular degrees.

It's not hard to imagine the sky as a sphere that measures 360 degrees — after all, space surrounds Earth on all sides. Standing in one spot on Earth, if you trace the sky from horizon to horizon, that would equal 180 degrees. Remember, the other 180 degrees is under the horizon.

If you want to measure the distances between two objects — say, between the Moon and Venus as they appear together in the sky — you can use your hand as a measuring tool. It all lies within your fingers.

Hold your hand at arm's length. The width of your pinky finger equals 1 degree. The width of your three middle fingers held at arm's length equals five degrees; a closed fist is 10 degrees; the distance between the tip of your index finger and the tip of your pinky is 15 degrees; and the distance between your thumb and pinky is roughly 25 degrees.

This is particularly helpful when trying to see how high or low something is above the horizon.

You can practice measuring the degrees with stars found in the Big Dipper. The chart shows how to find the Big Dipper using Polaris.

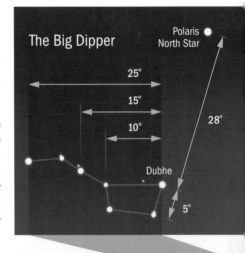

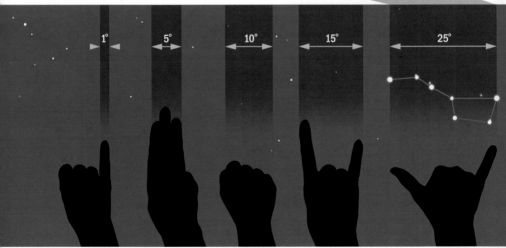

A view of Polaris and the Big Dipper over Dinosaur Provincial Park, Alberta

Binoculars and Telescopes

Using your eyes to move around the sky allows you to see the Moon, the planets, the Sun and even some objects outside of our Solar System. However, having binoculars or a telescope opens up the Universe to you.

Binoculars are an amazing first tool for seeking out the marvelous wonders of the night sky. A good pair of binoculars can reveal the intricacies of the Moon, showing its peaks and valleys, or uncover the daily motion of Jupiter's moons as they dance around our Solar System's largest planet. A decent telescope can make these objects appear bigger and show details of star clusters or nebulae (gas clouds in space).

But the question that often arises is what type of binoculars or telescope should I get?

For binoculars, it's important to understand two things: the **magnification** (or power) and the **aperture**. Binoculars are usually represented by two numbers, separated by an "×." Two example binocular numbers are 7×50 or 10×50. The magnification is the first number, and it represents the number of times larger something will appear compared to viewing it with the naked eye. The second number is the aperture in millimeters (fractions of an inch), and it represents the diameter of each lens. Good viewing in part depends on how well your binoculars can gather light. If your binoculars have a larger second number, they have a larger aperture and can gather more light from distant objects. But that comes with a cost. Bigger binoculars are heavier and, therefore, more difficult to hold, which means you will need a tripod to steady your view.

Another number often given for binoculars is the **field of view**, or FOV. This is how wide you will be able to see in degrees (see "Handy Sky Measures," on page 6). In general, the higher the magnification you use, the smaller your field of view will be. At times, this will mean you need to make a choice. Do you want to zoom in on a particular crater on the Moon, for example, or observe a more spread-out chain of mountains? Making these decisions is difficult for astronomers, too, so don't worry if it feels hard — it gets a bit easier to decide what to do with practice.

The top end of handheld binoculars is typically 7×50. Larger than that, it's best to get a tall tripod if you want the best view possible and to share views with others.

We recommend using binoculars for a few months before investing in a telescope. Once you are comfortable with binoculars, luckily, your knowledge will be useful because telescopes use many of the same definitions for observing.

Remember that light-gathering ability is important; a typical beginner's telescope usually ranges between 50 to 150 millimeters (2 to 6 inches) in aperture, the diameter of the main lens or mirror. The bigger the **aperture**, the more light the lens will gather, and the brighter the view will be. Invest in a sturdy tripod, too, so that the telescope will not shake when you use it.

The **magnification** of a telescope depends on the eyepiece used; magnification is calculated by dividing the focal length of the telescope by the focal length of the eyepiece (using like units). For example, a telescope with a focal length of 1,000 millimeters and a 10-millimeter eyepiece will magnify an object 100 times. The same 10-millimeter eyepiece in a 500-millimeter telescope would will magnify an object 50 times.

While a bigger aperture and higher magnification may seem like the best way to go, it's best not to get too caught up in choosing a telescope with the highest numbers. What is important is how you plan to use it. Since most people are unlikely to have backyard observatories, the most important thing may be portability. Too heavy a telescope means you're unlikely to haul it out to the backyard or to a vacation spot, far from city lights.

When it comes to telescopes, there are a wide array of choices. Some of the most popular are refractors, reflectors and compound telescopes. Each type has its own benefits, and it all depends on what you prefer. We therefore recommend you try testing out different telescopes at a local star party held by astronomy groups before making the investment. Alternatively, check reputable astronomy magazines or forums for recommendations for beginners; in most cases, all it takes is a little Internet searching and some patience.

A **refractor telescope** has a front lens that focuses light to form an image at the back, and an eyepiece that acts like a magnifying glass to allow your eye to focus on that image. Refractors tend to be more reliable, as their lenses are fixed in place and, therefore, don't get out of alignment as easily as some other types of telescopes. Refractors are best used for planetary and lunar observing as well as viewing double stars.

A **reflector telescope** uses a mirror to gather and focus light. The advantage for reflectors is they don't usually suffer from **chromatic aberrations**, when light of different wavelengths (i.e., colors) doesn't focus on the same point; this makes them ideal to observe distant objects like star clusters. Chromatic aberrations cause the different colors of the spectrum to split and the image to appear blurry, which isn't great when looking at a group of stars. A **Dobsonian telescope**, which is a variant of a reflector with a simple mount, gives you a far bigger aperture for far less the cost of other telescopes.

Compound or **catadioptric telescopes** combine lenses and mirrors to form an image. The **Schmidt-Cassegrain**, a type of compound telescope, has a compact design that makes it quite popular. With this instrument, an astronomer can get a bigger aperture in a smaller sized telescope. This telescope type is also portable, making it easier to move the equipment into remote areas.

Star parties are great opportunities to check out different telescopes and chat with fellow enthusiasts

Stars

Stars come in many different varieties. Our Sun is a **yellow dwarf star**, on the **main sequence**, meaning that it's converting hydrogen into helium at its core, like most other stars. When a star does this conversion, it releases a tremendous amount of energy. This energy, in the form of sunlight, allows life to thrive on Earth.

The Sun, at 4.5 billion years old, is considered middle-aged. When it dies, five billion years from now, it will at first swell, becoming a **red giant** and engulfing the inner planets, then it will slough off its outer layers to become a **white dwarf**.

One of the most interesting types of stars, some might argue, are **red supergiants**. These colossal stars — roughly 1,400 times the mass of the Sun — have relatively short lifespans.

And when supergiants do die, they do so in a spectacular fashion, in an explosion called a **supernova**. A supernova occurs when a star can no longer convert hydrogen into helium; eventually the core is converted into iron. During a supernova, the star first collapses and then explodes outward, creating even heavier elements.

Betelgeuse, a star found in the left shoulder of Orion, is a red supergiant. While many far-away and faint supernovae have been witnessed in modern history in other galaxies, none have occurred in our own galaxy — the **Milky Way**. When Betelgeuse goes supernova, its brightness will rival the full Moon in our sky: it will even be visible in daylight. If you're hoping to see Betelgeuse explode in your lifetime, you're not alone. However, estimates peg Betelgeuse's death for some time within the next 100,000 years.

The most common star in the Universe is a **red dwarf**, which is a cool star that's much smaller than the Sun. Astronomers often search for potentially habitable planets around these stars, because the inherent dimness and smaller size of red dwarfs makes it easier to spot planets. The stars, however, can be quite volatile, occasionally releasing a tremendous amount of radiation. Intense radiation is not a friendly process for most life forms.

The **Hertzsprung-Russell diagram** (H-R diagram) was developed in the early 1900s by Ejnar Hertzsprung and Henry Norris Russell, following the research findings by two Harvard computers, Annie Jump Cannon and Antonia Maury. The diagram plots the temperature of

Top 10 Brightest Stars in the Night Sky	
1.	Sirius
2.	Canopus
3.	Alpha Centauri
4.	Arcturus
5.	Vega
6.	Capella
7.	Rigel
8.	Procyon
9.	Achernar
10.	Betelgeuse

The Harvard Observatory Computers

In 1881, Edward Charles Pickering, the director of the Harvard Observatory, hired a team of women to compute and catalog photographs of the night sky. Their work was incredibly important in providing the foundations of astronomical theory. Annie Jump Cannon, one of the computers, added to work done by fellow computer Antonia Maury and developed a system of classifying stars that is still used today.

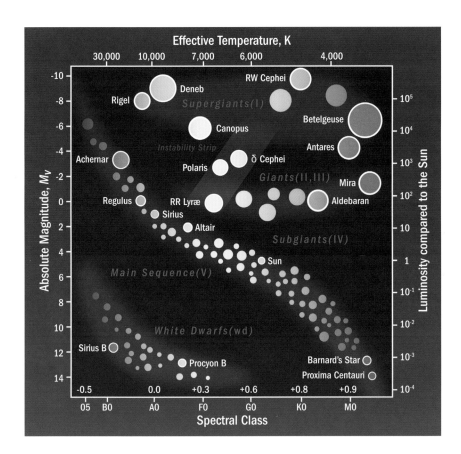

Effective Temperature, K

stars against their luminosity. Just as we do, stars go through certain stages in their lives. The H-R diagram provides astronomers with the information about a star's current age. Main-sequence stars that are fusing hydrogen into helium — such as our Sun — lie on the diagonal branch of the diagram.

Finally, there's a star's **magnitude**, or apparent brightness. Magnitude is measured on a scale where the higher the number, the fainter it is — and negative numbers are brighter than positive numbers. For example, Sirius, the brightest star in the night sky, measures –1.4 on this scale. Polaris is 2.0, and the Sun is

–27. We also use magnitude to measure the brightness of other celestial objects, like the Moon, planets, asteroids and comets.

Note there is a difference between apparent magnitude and absolute magnitude. Apparent magnitude is the brightness of an object that we observe from Earth, but absolute magnitude is the the brightness of an object if it were placed 32.6 light-years from Earth. This second measure helps astronomers directly compare the magnitudes of objects and is what is used for the H-R diagram. On this scale, Sirius has a magnitude of 1.4, Polaris is –3.6 and the Sun is 4.8.

Constellations

Constellations, a group of stars that make an imaginary image in the night sky, have been around since ancient times. Typically, the images astronomers use are based on Greek, Roman and Arabic mythologies. The International Astronomical Union (IAU) recognizes a total of 88 official constellations.

Some of the most recognizable constellations in the Northern Hemisphere are Orion (the Hunter), Cygnus (the Swan), Leo (the Lion), Gemini (the Twins), Scorpius (the Scorpion) and Ursa Major (the Great Bear).

More recently, there has been more effort to acknowledge constellations of Indigenous peoples. The naming of Indigenous constellations varies around the world, making these constellations regionally distinct. Some groups see

Ursa Major as a bear, or a caribou. The Cree see Corona Borealis as seven birds, and Cepheus as a turtle. To the Navajo, Polaris is Nahookos Biką, meaning central fire. For some Indigenous peoples, such as the Inuit, the Northern Lights are dancing spirits.

Aside from the constellations, there are also **asterisms**, a group of stars within a constellation (or sometimes from several different constellations) that forms its own distinct pattern. The Big Dipper is probably the most famous asterism, as its stars lie within the constellation of Ursa Major. There's also the Summer Triangle, with bright stars from Cygnus, Lyra (the Lyre) and Aquila (the Eagle), and the Winter Triangle with stars from Orion, Canis Major (the Great Dog) and Canis Minor (the Little Dog).

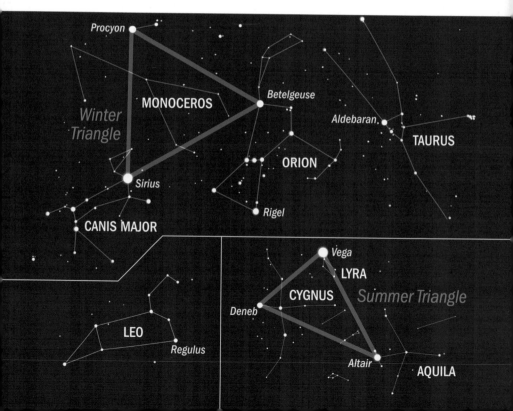

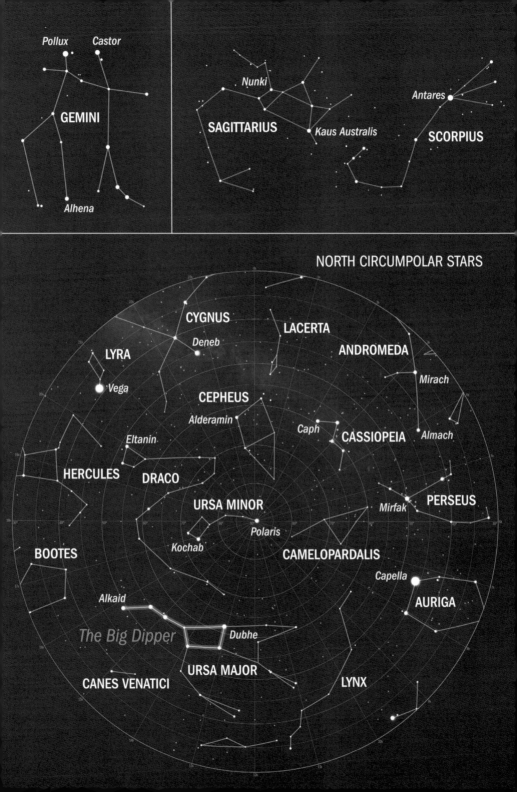

GEMINI

Pollux Castor

Alhena

SAGITTARIUS

Nunki

Kaus Australis

SCORPIUS

Antares

NORTH CIRCUMPOLAR STARS

CYGNUS

Deneb

LYRA

Vega

LACERTA

ANDROMEDA

Mirach

CEPHEUS

Alderamin

Caph

CASSIOPEIA

Almach

Eltanin

HERCULES

DRACO

URSA MINOR

Polaris

PERSEUS

Mirfak

BOOTES

Kochab

CAMELOPARDALIS

Capella

AURIGA

Alkaid

The Big Dipper

Dubhe

URSA MAJOR

CANES VENATICI

LYNX

Comets and Meteors

Comets are icy balls of debris — specifically, dust and ice — left over from the formation of our Solar System. They are sometimes referred to as "dirty snowballs" and can be stunning objects to see in the night sky when tails of dust and ionized gas fan out behind their cores. While we know of many comets, predicting the appearance of a bright one is all but impossible. Comets tend to fall apart before becoming too bright. The Sun's gravity and heat are strong, and comets themselves are very delicate visitors from the outer Solar System or beyond. Many comets literally crumble under the pressure as they dive in toward the Sun and the inner Solar System, where our planet resides.

Bright comets can be marvelous. On August 17, 2014, Comet Lovejoy C/2014 Q2 was spotted as it came toward the inner Solar System.

Comet Lovejoy C/2014 Q2

This comet produced a spectacular tail as it neared the Sun. Tails occur when ice **sublimates**, turning directly from a solid into a gas.

There have been other wonderful naked-eye comets that have graced our night sky, including Comet Hyakutake C/1996 B2, Hale-Bopp C/1995 O1, or Comet NEOWISE C/2020 F3. Each comet has the year of its discovery in its official name, so NEOWISE was found in 2020, Hyakutake in 1996, and so on. Many Northern Hemisphere observers were treated to a special show when Comet NEOWISE appeared in the skies in 2020. It was one of the brightest comets to appear in a generation — even city dwellers could spot it through light pollution.

Periodic comets, or ones that we can predict, are given the designation "P," while those that appear unexpectedly are given the designation "C." For example, Halley's Comet, or 1P/Halley, appears roughly every 76 years — a clear P. The last time it passed was 1986; the next time will be in 2061. In general, C comets are brighter than P comets because P comets have sublimated their material into space from repeated trips by the Sun.

A **meteor** (from the Greek *meteoros*, meaning "high in the air") is the light, heat and (occasionally) sound phenomena produced when a

Comet NEOWISE C/2020 F3

The Geminid meteor shower

meteoroid, or small debris left in space from comets or **asteroids** (space rocks), collides with molecules in Earth's upper atmosphere. We also call meteors **shooting stars**.

When a meteoroid enters Earth's atmosphere, the surface of the object is heated. Then, at a height typically between 119.8 and 79.9 kilometers (74.5 miles and 49.7 miles), the meteoroid begins to **ablate**, or lose mass. Meteoroid ablation usually happens through vaporization, although some melting and breaking apart can also occur.

Meteoroids are usually the size of a small pebble, but they range in size considerably; some can be as small as the size of the tip of a ballpoint pen, while more unusually, they can be several meters or kilometers across. If a meteoroid is large enough that it doesn't burn up entirely and reaches the ground, it is called a **meteorite** – and we have recorded instances of meteorites causing damage on Earth.

NASA keeps a sharp eye out for threatening objects in space and, so far, has found nothing of concern. We know a big meteorite strike could happen eventually, though. Just ask the dinosaurs, which were likely wiped out by a meteorite at least 11 kilometers (7 miles) across about 66 million years ago. Thankfully, such events are quite rare, happening every few million years on average.

Meteoroids can be divided into two groups: **stream** and **sporadic meteoroids**. Stream meteoroids have orbits around the Sun that often can be linked to a parent object – in most cases a comet, but it can sometimes be an asteroid. When Earth travels within the orbit of a stream, we get a **meteor shower**. Because all the objects (meteoroids) move in the same direction, they all seem to come from one point in the sky, called the **radiant**. The constellation containing the radiant (or in some cases a nearby star) is what gives a meteor its name, such as the Perseids or Geminids (two of the most active meteor showers).

Sporadic meteors just occur in random parts of the sky, with no radiant.

Meteor Showers in 2021

There is nothing more wonderful than stepping out and looking up at the night sky, only to see a brief streak of light flash against the stars.

Almost monthly, we get major meteor showers. Some showers are best seen from the Northern Hemisphere, while some are better visible in the south, depending on the radiant.

When astronomers talk about the "peak" of a meteor shower, they're referring to the **Zenithal Hourly Rate**, or the ZHR. This is the rate of meteors a shower would produce per hour under clear, dark skies and with the radiant at the zenith. In practice, a single observer will likely see considerably fewer meteors than the ZHR suggests, owing to the presence of moonlight, light pollution, the location of the radiant, poor night vision and inattentiveness.

Here is a list of major meteor showers that you can enjoy simply by staying warm and looking up. No special equipment is needed beyond your eyes; just make sure to give yourself about 20 minutes to adjust to the darkness before searching for meteors.

Quadrantids:
December 27, 2020–January 10, 2021
The Quadrantids might be one of the best meteor showers of the year, with a ZHR of 120 for a brief interval of time. The only thing holding it back from earning the title is that January tends to be cloudy over North America, and the shower's peak has a brief window of six hours, making the average hourly rate closer to 25. But there's good news: the meteors often produce bright **fireballs**, or bright meteors with a magnitude higher than –4.0. This shower was named after a constellation that no longer exists, Quadrans Muralis, but its radiant lies between Boötes and Draco. The bad news is this year there will be a waning gibbous Moon that will wash out all but the brightest meteors.
Parent Object: Asteroid 2003 EH1
2021 Peak Night: January 2–3

Lyrids: April 16–30, 2021
After a dearth of meteor showers in the months of February and March, April brings us the Lyrids. This shower produces a ZHR of 18, so it's not a particularly strong shower, but the meteors that do appear tend to produce fireballs. A waxing gibbous Moon will be up in the sky; however, viewing improves after the Moon sets around 02:00 local time.
Parent Object: Comet C/1861 G1 (Thatcher)
2021 Peak Night: April 21–22

Eta Aquariids: April 19–May 28, 2021
The Eta Aquariids aren't a stellar show for the Northern Hemisphere, since the radiant rises in early dawn, but they can still produce a ZHR of 20 meteors, if you care to get up early enough. Rather than fireballs, this shower tends to produce long trains through the sky. A waning crescent Moon will cause minimal interference during the peak.
Parent Object: Comet 1P/Halley
2021 Peak Night: May 5–6

Perseids: July 17–August 26, 2021
For those in the Northern Hemisphere, the Perseids are considered the best show of the year. The weather is warm and there are fewer clouds at this time of year. The shower's ZHR is 100, though rates of 50 to 75 are more commonly seen on the peak night. A waxing crescent Moon will be in the sky on the peak night, but it will set shortly after dusk.
Parent Object: Comet 109P/Swift-Tuttle
2021 Peak Night: August 12–13

Southern Taurids:
September 10–November 20, 2021
The Southern Taurids last for two months and have several minor peaks. Though this shower produces a ZHR of only 5, it can produce some fireballs. The shower is stronger in the Southern Hemisphere, though we can also catch a few meteors in the north. The Moon that night will be just after first quarter but will have set by nightfall.
Parent Object: Comet 2P/Encke
2021 Peak Night: October 9–10

Orionids: October 2–November 7, 2021
This shower is considered medium strength, though it can sometimes surprise us with more activity. On average, the Orionids produce a ZHR of 10 to 20 meteors, though from 2006 to 2009, the shower produced roughly 50 to 75 an hour. The meteors that enter our atmosphere are fast — roughly 66 kilometers (41 miles) per second — and, as a result, produce long, glowing trains. Fireballs are also possible. Unfortunately, a full Moon will be in the sky all night long during the peak, which means observers can expect to see only five meteors per hour.
Parent Object: Comet 1P/Halley
2021 Peak Night: October 20–21

Northern Taurids:
October 20–December 10, 2021
Like the Southern Taurids, the Northern Taurids last roughly two months and have a ZHR of 5. There are often reports of an increase of fireballs during the period when the two showers overlap. The peak occurs alongside a first quarter Moon that sets near midnight.
Parent Object: Comet 2P/Encke
2021 Peak Night: November 11–12

Leonids: November 6–30, 2021
While the Leonids don't produce a high number of meteors an hour (they have a ZHR of 15), they can produce outbursts, or meteor storms; the most recent such outbursts occurred in 1999 and 2001. Unfortunately, the next outburst isn't expected until 2099. Still, the Leonids do put on a show, with bright meteors and long-lasting trains. Peak night comes just two days before the full Moon, which will make this shower difficult to observe.
Parent Object: Comet 55P/Tempel-Tuttle
2021 Peak Night: November 17–18

Geminids: December 4–17, 2021
Due to the weather — with increasing cloudiness and chillier temperatures — December isn't an ideal time to catch a meteor shower. But December is the month with the most active meteor shower of the year, called the Geminids. This shower has a ZHR of 150, and the meteors it produces are usually bright and colorful. A waxing gibbous Moon provides a challenge for evening observing, though it will set close to the time the radiant is high in the sky. Then observers can expect to see a meteor a minute.
Parent Object: Asteroid 3200 Phaethon
2021 Peak Night: December 13–14

Ursids: December 17–26, 2021
Right on the heels of the Geminids is the often-forgotten Ursid meteor shower. The ZHR for this shower is 10, but sometimes you might catch an outburst that could produce a ZHR of 25. The peak night comes just two nights after a full Moon, which means it will be difficult for observers to catch any activity.
Parent Object: Comet 8P/Tuttle
2021 Peak Night: December 21–22

The Moon

The Moon is most often the first astronomical object to capture anyone's attention in the night sky. It's also likely the first object most people see up close, whether it be through binoculars or a telescope.

Our nearest celestial neighbor has been an object of fascination since humans first looked up to the sky. Early scientists in ancient observatories carefully tracked the cycles of the Moon. Some 23 centuries ago, Aristarchus of Samos carefully observed a total lunar eclipse and derived impressively accurate measurements of the Moon's diameter. He estimated its diameter was one-third that of Earth. He was close: the actual percentage is 27.2, and its diameter is roughly 3,540 kilometers (2,200 miles).

It takes about 29.5 days for the Moon to go through its cycle of **phases**. When you can't see the Moon at all, it's called a **new Moon**.

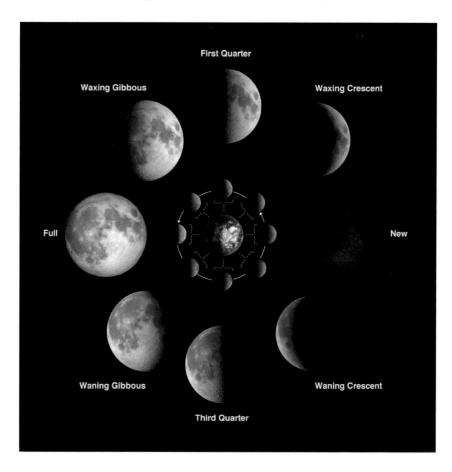

First Quarter

Waxing Gibbous

Waxing Crescent

Full

New

Waning Gibbous

Waning Crescent

Third Quarter

When the Moon is fully illuminated, it is a **full Moon**. The list of phases is new Moon, waxing crescent Moon (when the Moon is just a thin crescent), first quarter Moon (when it appears half full), waxing gibbous Moon (when the Moon is between half-full and full), then full Moon — before the process runs in reverse from full to new: waning gibbous Moon, third or last quarter Moon, waning crescent Moon, new Moon.

The Moon is tidally locked with Earth, meaning that we only ever get to see one face of it. The Sun does shine on the other side of the Moon when we can't see it, so don't call the far side of the Moon the dark side.

While it may seem that we always see the same Moon features month after month, that's not entirely true. The Moon oscillates from our perspective, a process called **lunar libration**. As a result, every so often we see some small part of the Moon that we don't always see.

Observing the Moon

You don't need a telescope to enjoy the Moon. If observed even briefly on a regular basis, Earth's only natural satellite has much to offer the naked-eye observer. Keep an eye on it, and you will notice things like its wandering path through the constellations, the changing phases, frequent **conjunctions** (where two objects appear close together in the sky) with planets or bright stars, occasional eclipses, lunar libration, earthshine (illumination on the Moon reflected from our planet) and other wonderful atmospheric effects.

If you happen to take a look through a telescope, binoculars or even a camera, our closest astronomical target offers an amazing amount of interesting detail. There are countless features on the lunar nearside, and more than 1,000 have been formally named by the IAU. Many of the greatest names in astronomy, exploration and discovery are commemorated through these names. For extra dramatic effect, try looking at a lunar feature when the **terminator** — the line of darkness and sunlight — runs nearby the feature. The extra shadow could make it easier to see details in the object you're interested in looking at.

An estimated three to four billion years ago, our Solar System was bombarded by debris for hundreds of millions of years. This period is called the Late Heavy Bombardment, and you can see many of the scars from that time left over on the Moon.

Because the Moon has almost no atmosphere and no tectonic activity, those scars have remained over billions of years. Some of that activity created deep **impact basins** that filled with lava flows and hardened into basaltic rock. These largely circular regions were termed **maria**, or seas, by early astronomers because they resembled Earth's oceans.

The brighter white areas, which are highlands that surround the maria and dominate the southern portion of the Earth-facing hemisphere, feature many ancient **impact craters**.

After the bombardment slowed to a trickle, the Moon remained relatively free from intense impact activity — meaning the Moon has changed very little since billions of years ago. But, just like on Earth, meteorites still periodically slam into the Moon today.

The Moon map on the next two pages shows a few interesting features for you to target as you observe the Moon.

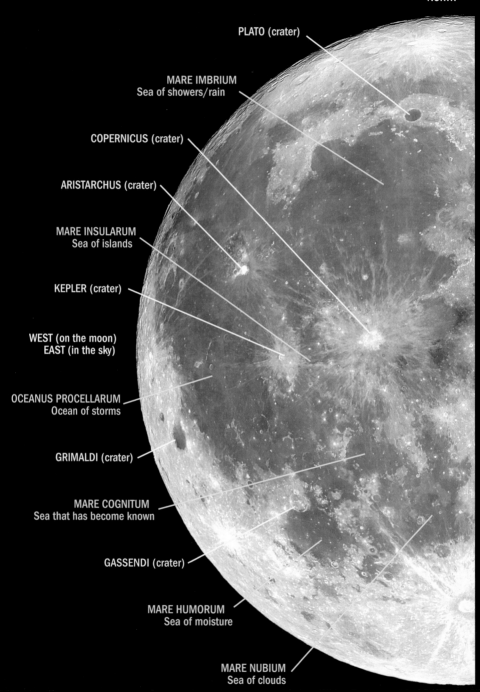

PLATO (crater)

MARE IMBRIUM
Sea of showers/rain

COPERNICUS (crater)

ARISTARCHUS (crater)

MARE INSULARUM
Sea of islands

KEPLER (crater)

WEST (on the moon)
EAST (in the sky)

OCEANUS PROCELLARUM
Ocean of storms

GRIMALDI (crater)

MARE COGNITUM
Sea that has become known

GASSENDI (crater)

MARE HUMORUM
Sea of moisture

MARE NUBIUM
Sea of clouds

SOUTH

MARE FRIGORIS
Sea of cold

MARE SERENITATIS
Sea of serenity

MARE TRANQUILLITATIS
Sea of tranquility

MARE CRISIUM
Sea of crises

MARE FECUNDITATIS
Sea of fecundity/
fertility

EAST (on the moon)
WEST (in the sky)

LANGRENUS (crater)

MARE NECTARIS
Sea of nectar

STEVINUS (crater)

TYCHO (crater)

The Sun

Our Sun highly influences Earth. It is connected with our seasons, weather, ocean currents and climate, and its power makes life itself possible.

Unlike Earth, the Sun isn't a solid body, and due to that, different parts of it rotate at different rates. At the equator, for example, it spins roughly once every 25 days.

In the core of the Sun, where hydrogen atoms fuse to make helium, temperatures are a searing 15 million degrees Celsius (27 million degrees Fahrenheit). The tremendous amount of energy generated at the core — **thermonuclear fusion** — is carried outward by radiation, taking roughly 170,000 years to get from the core to the top of the convective

Take Care of Your Eyesight

Do not observe the Sun without special protective equipment, such as a certified solar filter that covers your eyes or telescope aperture entirely. Unprotected observations even for a few seconds could damage your eyesight permanently.

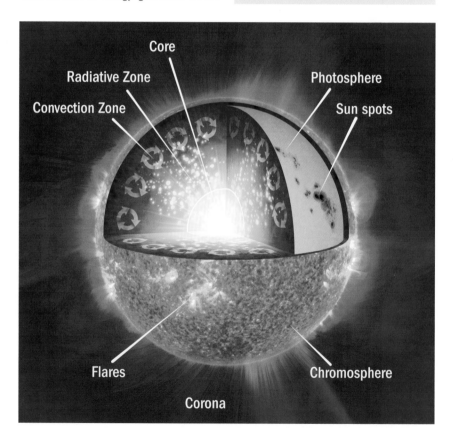

Core

Radiative Zone

Convection Zone

Photosphere

Sun spots

Flares

Chromosphere

Corona

zone. As hot as the core is, at the surface, it's a much different story – the temperature is a much "cooler" 5,500 degrees Celsius (10,000 degrees Fahrenheit).

The Sun also has a roughly 11-year cycle, that has a **maximum** (a period of increased solar activity) and a **minimum** (a period of low solar activity). During a maximum, sunspots, which are cooler regions on the surface of the Sun, increase. Sometimes, the magnetic field lines of these sunspots can become entangled, finally snapping and releasing a tremendous amount of radiation into space. This process is called a **solar flare**. And often a **coronal mass ejection (CME)** can follow a flare, with charged particles of the Sun speeding outward into space. If these particles reach Earth, they can disrupt radio transmissions, damage satellites and, more positively, interact with our magnetic field. When the particles do interact

A New Solar Cycle

Right now we are just starting Solar Cycle 25. After a quiet solar minimum, we can now expect to see more solar activity. Solar Cycle 24 was the weakest cycle in the last 200 years.

with the field, they can produce the beautiful Northern Lights or aurora borealis.

Though beautiful, it must be said that these outbursts from the Sun can also cause power outages, as was experienced in Quebec in 1989. As such, astronomers are keen to better understand our nearest star using many spacecraft – such as the Parker Solar Probe, which was launched in 2018. Keep watching the eight-year adventure of the Parker Solar Probe as it studies the Sun's activity.

The Northern Lights

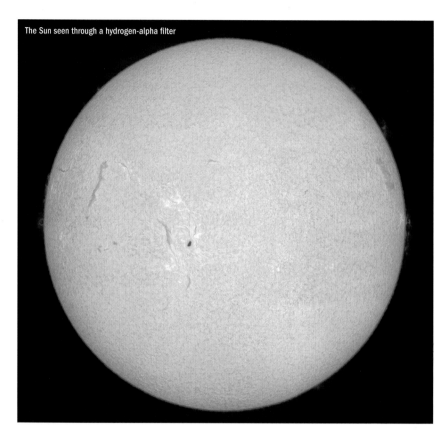
The Sun seen through a hydrogen-alpha filter

Observing the Sun

The Sun, our closest star, is a wonderful object to observe with any type of telescope, if used safely. The only safe filter is the kind that covers the full aperture of the telescope, at the front, allowing only 1/100,000th of the sunlight through the telescope. Without this filter, you risk permanently damaging your telescope or, far worse, your eyes. Ensure that the filter material is specifically certified as suitable for solar observing, and *take great caution every time you observe the Sun.*

The preferred solar filters are made of Thousand Oaks glass or Baader film. Thousand Oaks glass gives the Sun a golden-orange color, while Baader film gives the Sun a more natural white look, which can provide good contrast if there are any bright spots, or **plages**, on the Sun. Other solar-filter options include lightweight Mylar film (which gives the Sun a blue-white tint), and eyepiece projection. Projecting the image from the eyepiece onto white cardstock paper, or a wall, produces an image that can be safely shared with others.

While it might seem like the Sun is just this boring yellow ball in the sky, there are many things to see on its surface, including **prominences**, **filaments**, flares and sunspots. These phenomena result from the strong magnetism within the Sun, which erupts to the surface.

Prominences are solar plasma ejections, some of which fall back to the Sun in the form of teardrops or loops; against the Sun's disk, prominences viewed from above appear as dark filaments. Sunspots are cooler regions of the Sun that appear dark on the face of the surrounding surface. If the highly magnetic lines of sunspots intertwine, they can snap and cause a solar flare — a bright, sudden eruption of energy that can last from a few minutes to several hours. It should be noted that to see prominences and some other solar features, you need very specialized filters.

If you do not have the proper equipment to observe the Sun, you can always visit the Solar and Heliospheric Observatory (SOHO) and the Solar Dynamics Observatory (SDO) websites (see the list of resources on page 118) to see daily images of the Sun.

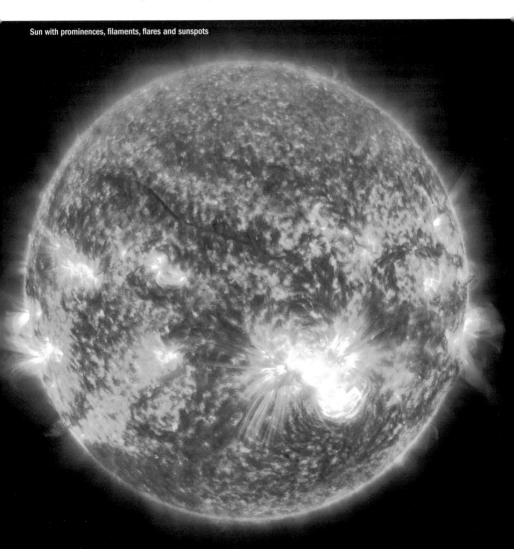

Sun with prominences, filaments, flares and sunspots

Eclipses

Eclipses — whether they're solar or lunar — are seemingly magical occurrences. In fact, that's exactly what many ancient civilizations believed. Several legends suggested a celestial being devoured the Sun during a solar eclipse. For the Chinese, that being was either a dog or a dragon. The Vikings believed it was two wolves called Hati and Skoll. For the Vietnamese, it was a giant frog. Today, we understand that eclipses are awe-inspiring celestial events resulting from the movement and positions of the Earth, Moon and Sun.

Solar eclipse

A solar eclipse is a truly chance occurrence. The Sun's diameter is roughly 400 times that of the Moon, but the Sun is also about 400 times farther away from Earth than the Moon is from Earth. This means the Moon and the Sun appear roughly the same size in the sky, with only about half a degree across of difference. In a **total solar eclipse**, the Moon covers the entire face of the Sun, revealing the Sun's stunning corona and prominences. As totality begins and ends, sunlight peeking around the mountainous limb of the Moon creates fleeting effects such as the Diamond Ring and Baily's Beads. If the Moon were farther away or smaller, we wouldn't get this marvelous sight. In fact, the elliptical nature of both the Earth's orbit around the Sun and the Moon's orbit around the Earth means that the apparent diameters of both objects vary significantly. Consequently, it is common for eclipse durations to vary by several minutes.

It should be noted that the totality only occurs in a very narrow path. Outside of this path, observers will only see a partial eclipse. That's why solar eclipse enthusiasts will travel great distances to take in the captivating sight of a totality. (Remember to never look at a solar eclipse unless you use special equipment recommended by The Royal Astronomical Society of Canada or the American Astronomical Society.)

Because the Moon orbits at a 5-degree inclination to the ecliptic, we don't get a solar eclipse every month, and sometimes we get an **annular solar eclipse**, when the Moon doesn't entirely cover the Sun's disk. There are two, sometimes three, total or annular eclipses in a given year. There may also be eclipses where the Moon's central shadow entirely misses the Earth but the Sun is partially eclipsed.

As for lunar eclipses, this occurs when Earth is situated directly between the Sun and the Moon. The Moon drifts through Earth's two shadows: the **penumbra**, which is the fainter, outer shadow, and the deeper shadow that is called the **umbra**. A **penumbral eclipse** is difficult to see with the naked eye as the brightness of the Moon doesn't appear to dim much. But a **partial** or **total lunar eclipse**, when the Moon passes through the umbra, is much more dramatic. During a total lunar eclipse, the Moon can turn an orange-red color, depending on Earth's atmosphere.

While not all eclipses will be visible from North America, you can watch online on sites like SLOOH or The Virtual Telescope Project. You can also see astronomer Fred Espenak's webpage eclipsewise.com for a comprehensive treatment of solar and lunar eclipses.

Eclipses in 2021

In 2021, there are no total solar or lunar eclipses completely visible from North America (other than from Hawaii and parts of Alaska).

Lunar Eclipses

Total Lunar Eclipse: May 26, 2021

The first eclipse of the year occurs on May 26. This total lunar eclipse can be seen from parts of the Americas, the Pacific, Australia and eastern Asia.

Unfortunately, this eclipse is only partially visible across Canada and the U.S., with the west being the best location. Only Hawaii will be able to see the entire eclipse through totality. Europe, Africa and western Asia are completely shut out.

Eclipse times (UTC)	
Moon enters the penumbra (P1)	08:47
Partial eclipse begins (U1)	09:44
Totality begins (U2)	11:11
Greatest eclipse	11:18
Totality ends (U3)	11:25
Partial eclipse ends (U4)	12:52
Moon exits penumbra	12:49

The eclipse occurs 0.4 days after the Moon is at **perigee**, or at its closest to Earth in its orbit, and is a very shallow eclipse as the Moon just skirts the umbra. As a result, the totality lasts roughly 14.5 minutes.

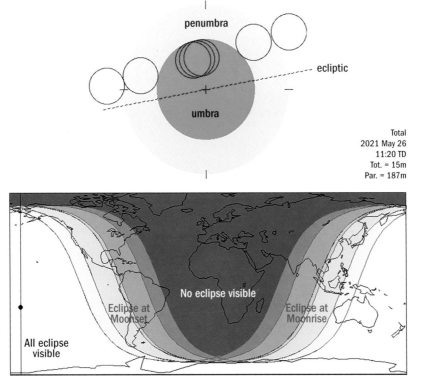

penumbra

ecliptic

umbra

Total
2021 May 26
11:20 TD
Tot. = 15m
Par. = 187m

No eclipse visible

Eclipse at Moonset

Eclipse at Moonrise

All eclipse visible

Thousand Year Canon of Lunar Eclipses © 2014 by Fred Espenak

Partial Lunar Eclipse: November 19, 2021
This time, North America is in a prime location to enjoy this eclipse. Though this eclipse is considered partial, most of the Moon will be in the umbra, making this a deep partial eclipse. On the east coast of Canada and the U.S., the partial eclipse may still be underway at sunrise.

Eclipse times (UTC)	
Moon enters the penumbra (P1)	06:02
Partial eclipse begins (U1)	07:18
Greatest eclipse	09:02
Partial eclipse ends (U4)	10:47
Moon exits penumbra	12:03

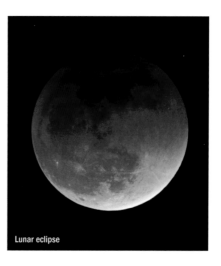
Lunar eclipse

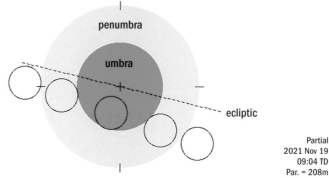

penumbra

umbra

ecliptic

Partial
2021 Nov 19
09:04 TD
Par. = 208m

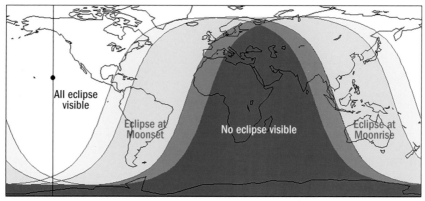

All eclipse visible

Eclipse at Moonset

No eclipse visible

Eclipse at Moonrise

Thousand Year Canon of Lunar Eclipses © 2014 by Fred Espenak

Solar Eclipses

Annular Solar Eclipse: June 10, 2021

Fifteen days after the total lunar eclipse, there's an annular solar eclipse. Though the eclipse will take place over North America, it will only be seen in northern Canada, Greenland and Russia. In other parts of the Northern Hemisphere, including northeastern North America, northern Europe and Asia, it will be seen as a partial solar eclipse.

The eclipse occurs 2.3 days after **apogee**, when the Moon is at its most distant in its monthly orbit around Earth, and it will last 3 minutes and 51 seconds.

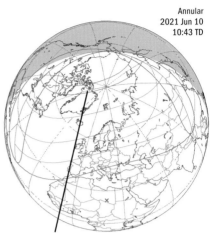

Annular
2021 Jun 10
10:43 TD

point of greatest eclipse

Thousand Year Canon of Solar Eclipses
© 2014 by Fred Espenak

Total Solar Eclipse: December 4, 2021

The last eclipse of the year is a total solar eclipse. However, this one occurs over Antarctica, with a partial eclipse occurring over other parts of the continent as well as South Africa and the southern Atlantic and Pacific Oceans. The eclipse occurs just 0.1 days before the Moon is at perigee.

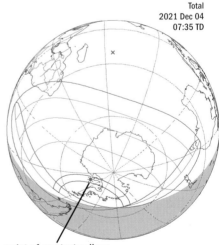

Total
2021 Dec 04
07:35 TD

point of greatest eclipse

Thousand Year Canon of Solar Eclipses
© 2014 by Fred Espenak

The Northern Lights

The Sun appears to be a bright, unchanging orb in the sky. However, the Sun is anything but unchanging: As we know, it is a ball of constant activity, and that activity has a large influence here on Earth.

One such activity is a solar flare, which is often followed by a coronal mass ejection (CME) that sends particles speeding along the solar wind. Solar flares can reach Earth and disrupt radio transmissions. If Earth is in the path of a CME, the particles can travel down our magnetic field lines toward the poles; this creates the beautiful Northern and Southern Lights, or **aurora borealis** and **aurora australis**, respectively.

As mentioned earlier, the Sun goes through an average 11-year cycle of activity during which it experiences a solar minimum and a solar maximum. In 2021, we will have just begun a new Solar Cycle, and as the cycle continues, we can expect to see more of the Northern Lights. It's worth noting that over the past few cycles the Sun has been less active than in the past.

Auroras come in different shapes and sizes. They can be steady, moving or rapidly pulsating. They also come in an array of colors, depending on how the particles interact with molecules at different altitudes. They are also hard to predict and can be difficult to see. Catching them is a special treat, even for experienced astronomers. Be aware that to the unaided eye the auroral colors are usually muted, as the light is not strong enough to stimulate our color vision. The bright colors seen in photographic reproductions of auroras show up when time exposures are used.

Green is the most common color and occurs when particles interact with oxygen molecules at an altitude of roughly 100 to 300 kilometers (62 to 186 miles). Between 300 to 400 kilometers (186 to 248 miles) the oxygen molecules produce a red color, and below 100 kilometers (62 miles) the molecules interact with nitrogen — producing a pink color.

While the interaction of these solar particles can produce a beautiful light show, they can also be destructive.

One of the most powerful events from a CME was called the Carrington Event. In 1859, English astronomers Richard Carrington and Richard Hodgson were the first to ever witness a solar flare. But when the particles reached Earth, the particles traveled along telegraph wires, in some reports even setting equipment on fire. The Northern Lights were visible as far south as Honolulu and the Southern Lights as far north as Santiago, Chile.

A massive CME, like the one that caused the Carrington Event, has a real chance of disrupting GPS and satellite communications and destroying electrical grids. As a result, space agencies have been launching more satellites and probes to the Sun to monitor space weather, and power companies have been developing contingency plans to ensure the next powerful solar eruption doesn't cause such damage.

The Northern Lights over Saskatchewan

The Planets

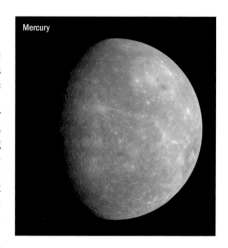

Mercury

As the planets (including Earth) revolve around the Sun, there are a number of notable events that take place, each of which has a specific name and meaning.

The inner planets Mercury and Venus never stray far from the Sun. When either planet is farthest from the Sun in the western evening sky, it is said to be at **greatest eastern elongation**; when either is farthest from the Sun in the eastern morning sky, it is said to be at **greatest western elongation**. These are the best opportunities to view the inner planets, though Venus can be seen at other times.

The term **conjunction** has a specific technical meaning — that is, when two objects have the same right ascension — but amateur astronomers also use the term casually when there is a close approach of two or more celestial objects.

Then there are **oppositions**: when the outer planets (Mars, Jupiter, Saturn, Uranus, Neptune) are opposite the Sun in the sky in right ascension. Around opposition dates, the outer planets appear closer, brighter and larger in diameter — great for telescopic viewing. The Sky Month-by-Month section (pages 40–113) lists eastern and western elongations, conjunctions and oppositions throughout 2021.

Mercury is a place of extremes. As the smallest of our Solar System's eight planets at one-third the size of Earth, Mercury is the closest planet to the Sun, sitting an average distance of 58 million kilometers (36 million miles) away. The planet, however, is only the second-hottest one in our Solar System. Mercury also orbits the Sun faster than any other planet and has the longest solar day, lasting 176 Earth days. Mercury has no moon.

The planet's surface warms to 427 degrees Celsius (801 degrees Fahrenheit) at the peak of its "noonday" heat; on the far side, its temperature drops down to a chilly –163 degrees Celsius (–261 degrees Fahrenheit). Mercury's coldest locations are deep in the shadows of the craters, where the Mercury Surface, Space Environment, Geochemistry and Ranging (MESSENGER) spacecraft mission to the small planet discovered ice. Temperatures in crater shadows are at –183 degrees Celsius (–297 degrees Fahrenheit).

Due to the planet's proximity to the Sun, Mercury is a challenging target to observe. When it is visible in Earth's sky, it appears in a narrow window of time in the dawn twilight just before sunrise or in the evening twilight just after sunset, depending on whether it is west or east of the Sun in the sky. To observe Mercury, one needs a low horizon unobstructed by trees or buildings, as it typically only 15 to 25 degrees away from the Sun.

Venus, the brightest planet in the sky, is a spectacular sight to see. Known as both the "morning star" and the "evening star," Venus is similar in size to Earth. But it's definitely not a place you'd like to visit. The planet is covered in dense clouds and, as a result, suffers from a runaway greenhouse effect. The planet's surface has temperatures of 460 degrees Celsius (860 degrees Fahrenheit) almost constantly.

Venus spins in the opposite direction than most planets, but very slowly. A day on Venus is longer than its year: one day takes 243 Earth days, while one year is 225 Earth days. Its orbit is, on average, 108 million kilometers (67 million miles) from the Sun. Being shrouded in white clouds, there is nothing much to see on Venus, but binoculars and small telescopes reveal that Venus passes through phases, similar to the Moon. (Mercury also shows phases, but they are more difficult to see, as the planet's disk appears so small.)

In 2021, you can see Venus in the morning sky beginning in January. On the morning of January 11, Venus is visible very close to the Moon, very low in the southeast in the morning twilight. Venus will have several more conjunctions with the Moon and planets throughout 2021.

There is no other planet that fascinates humans as much as **Mars**. In 1894, American astronomer Percival Lowell thought he could see deep canals carved through its surface, canals that were evidence of intelligent life. Later observations with spacecraft saw no evidence of these canals.

Mars is the most explored planet in our Solar System, and some even like to joke that it's the only planet with an entirely robotic "civilization" of Earth-made machines.

The orbiters, landers and rovers dispatched to the Red Planet have taught us a lot about its ancient past. Though rocky, dusty and seemingly devoid of any life, some evidence suggests Mars once had an ocean of water covering most of its northern hemisphere. Scientists sometimes cite the existence of water to suggest the planet was once habitable, though we have yet to find proof that life actually ever thrived there.

Mars is about half the size of Earth, with a day very close to Earth's 24-hour day. It lies roughly 228 million kilometers (142 million miles) away

from the Sun. Small telescopes (with a 70 mm diameter and below) will show the red disk of Mars; a larger, good-quality telescope at high magnification will show surface features such as ice caps, dark basins and light deserts.

NASA's Sojourner Rover roams the rocky terrain on Mars

In 2018 Mars was the closest to Earth (also meaning it was at its brightest in Earth's sky) than it had been since 2003. In 2021, the planet will shine brightly at the beginning of the year, and then will dim somewhat over the following six months, after which it gets too close to the Sun for observing. Oppositions of Mars occur about every 2 years, 2 months.

October 2020 saw a good opposition; the next opportunity will be in December 2022.

Jupiter is the king of our planetary system, at 11 times wider than Earth. This gas giant boasts the most spectacular storm formation in our Solar System, called the Great Red Spot. Jupiter orbits 772 million kilometers (479 million miles) from the Sun and a single day is about 10 Earth hours long. It orbits the Sun in about 12 Earth years.

Jupiter is the second-brightest planet in our night sky and is also one of the most enjoyable planets to observe. With a modest telescope, you can easily see the cloud bands in its atmosphere. But you can even enjoy the planet through a pair of binoculars; if you watch the planet every night, four of its 79 confirmed moons — Ganymede, Io, Europa and Callisto — can be seen changing positions night after night.

The planet is also home to the Great Red Spot (GRS), a massive, swirling egg-shaped storm. But that storm has been shrinking for reasons that are unclear to astronomers. In the 1800s, the GRS was estimated at 41,000 kilometers (25,476 miles) along its long axis. NASA's Juno spacecraft measured the GRS at 16,350 kilometers (10,159 miles) in width on April 3, 2017. Several astronomy apps predict the best time for viewing the GRS in a telescope on any night.

Jupiter spends all of 2021 in the adjacent constellations of Capricornus and Aquarius and is in opposition on August 19 at magnitude -2.9. One beautiful conjunction that is worth getting up for is Jupiter pairing with Mercury on March 5. The two planets will rise 50 minutes before the Sun in bright twilight, and they will be roughly one-third of a degree apart. Saturn will also lie to the right.

Speaking of which, who doesn't love **Saturn**? With its magnificent rings, the planet is often considered the jewel of our Solar System. With a pair of binoculars, Saturn's pancake shape is evident. Look through a telescope, even a modest one, and its fine rings are clearly visible.

Saturn is nine times wider than Earth and 1.4 billion kilometers (886 million miles) away from the Sun. This gaseous planet has the second-shortest day in the Solar System, lasting roughly only 10.7 hours, while its orbit takes 29.4 Earth years. The planet also has more than 80 moons, though its largest, Titan, is the only one easily viewed through a modest telescope.

Saturn is the second-largest planet in the Solar System. It's not the only planet to have rings (Jupiter, Uranus and Neptune all have ring systems), but it has by far the most spectacular ring system we can see. Saturn's rings are made up of billions of pieces of ice that range from tiny dust-sized grains to chunks as big as a house. The system extends roughly 282,000 kilometers (175,000 miles) from the planet but only about 9 meters (30 feet) vertically. There are several rings, mostly close to each other, but the main rings are called A, B, and C. The gap between the outer

Jupiter

Saturn

ring (A) and the first inner ring (B) is called the Cassini Division, following the discovery by Italian astronomer Giovanni Cassini.

Saturn spends all of 2021 in Capricornus and is at opposition on August 2 at magnitude +0.2.

Uranus is the faintest of the planets visible to the naked eye (in dark-sky conditions), and the third-largest planet overall. This giant ice planet has 13 faint rings and 27 small moons. But most interestingly, Uranus rotates on its side, at nearly a 90-degree angle.

In 1781, Uranus was the first planet to be discovered with the aid of a telescope by astronomer William Herschel. However, Herschel first believed it was either a star or a comet. It was confirmed as a planet two years later by Johann Elert Bode.

The planet is four times the size of Earth. A day on Uranus takes roughly 17 Earth hours, with one orbit taking 84 Earth years. Uranus orbits on average 2.9 billion kilometers (1.8 billion miles) from the Sun.

Uranus spends all of 2021 in Aries, reaching opposition on November 5 at magnitude +5.6. Although Uranus can be detected by the naked eye in ideal conditions, it is helpful to use binoculars to identify it among the surrounding stars with the help of a finder chart.

At an average distance of 4.5 billion kilometers (2.8 billion miles) from the Sun, **Neptune** is the farthest planet in our Solar System, taking about 165 Earth years to complete one orbit. Like Uranus, Neptune is roughly four times larger than Earth, and it, too, has a faint ring system.

Neptune also holds the distinction as being the windiest planet in our Solar System, with clouds of frozen methane being whipped across the planet at 2,000 kilometers (1,200 miles) per hour. A day on Neptune is about 16 Earth hours long.

Neptune spends all of 2021 in Aquarius, reaching opposition on September 14 at magnitude +7.8. Like Uranus, you can find Neptune through binoculars among the surrounding stars with the help of a finder chart.

Deep Sky Objects

The Universe has many amazing sights to behold, including stunning clusters of stars, nebulae and swirling galaxies.

There are two main types of star clusters: **open** and **globular**. Globular clusters are old star systems at the edge of spiral galaxies that can contain anywhere from thousands to millions of stars, packed in a close, roughly spherical form and held together by gravity.

Two beautiful globular clusters you can see from the Northern Hemisphere include Messier 13, found in the constellation Hercules (a hero from Greek mythology), and Messier 3 in the constellation Canes Venatici (the hunting dogs).

Open clusters contain anything from a dozen to hundreds of stars, but the stars are more spread out and found on the **galactic plane**, the plane on which most of a galaxy's mass lies. Perhaps the most famous open star cluster, the Pleiades (Messier 45) can be spotted with the naked eye and through light pollution.

As well, there are **nebulae** — clouds of dust and gas left over from a **supernova**, an exploding star. These are considered "stellar nurseries," as eventually the gas and dust will coalesce into new stars and potential stellar systems with planets, moons and, possibly, life. One of the most famous and easily visible is the Orion Nebula (Messier 42), found in the winter constellation Orion.

There are also **emission nebulae**, which are clouds of interstellar gas excited by nearby stars that emit their own light at optical wavelengths. **Planetary nebulae** are cloudy remnants leftover from stars that shed their gas and dust late in their lives. And finally, there are **dark nebulae**, interstellar clouds that are so dense that they obscure the light of the objects behind them.

Messier 13, a globular cluster

Messier Catalog

The Messier Catalog is a register of 110 objects in the night sky, including open and globular clusters, nebulae, galaxies and double stars. The catalog was started by Charles Messier in the 18th century. Messier was chiefly interested in finding comets, and the catalog is his list of non-comet objects that he observed. There are other objects that aren't on the Messier Catalog but are listed in the New General Catalog (NGC).

The Pleiades (Messier 45)

The Orion Nebula (Messier 42)

Galaxies

Galaxies come in many different shapes and sizes. There are four main types, however: elliptical, spiral, barred spiral and irregular. The diagram below shows Edwin Hubble's scheme for classifying galaxies, colloquially known as the "tuning fork" because of its shape.

Elliptical galaxies seem somewhat disorganized and look roughly egg-shaped. Shapes range from almost circular (E0) to very elliptical (E7).

Spiral galaxies have both a large central bulge and a thin disc of stars. Much like elliptical galaxies, these also range from tight spirals (Sa) to more diffuse (Sd).

Barred spiral galaxies are similar to spirals, but they have visible "arms" or bars near the center, and they range from tight (SBa) to diffuse (SBd).

There are also disc galaxies that don't have spiral arms. These are called **lenticular (lens-shaped) galaxies**. They are classified as S0.

The most common galaxy is the spiral, accounting for more than 75 percent of galaxies in the visible Universe. Our galaxy, the Milky Way, is believed to be a barred spiral. Our closest neighboring spiral galaxy is the Andromeda Galaxy (Messier 31), which is easily visible through binoculars or with the naked eye in dark skies.

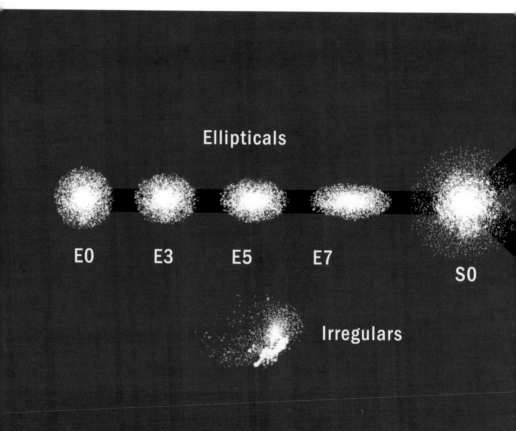

Ellipticals

E0   E3   E5   E7

S0

Irregulars

The Andromeda Galaxy (Messier 31)

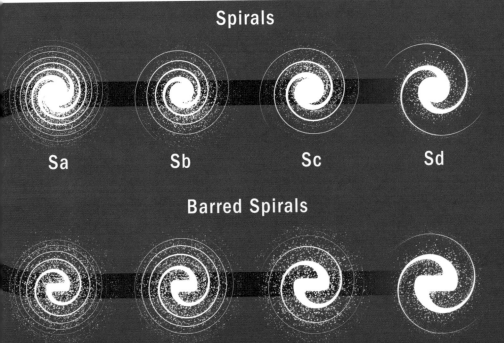

Spirals

Sa   Sb   Sc   Sd

Barred Spirals

SBa   SBb   SBc   SBd

The Sky Month-by-Month

Introduction

The following pages are your guide to the sky for every month in 2021. Each month features a calendar of events, information about Moon phases, sky charts facing south and north and descriptions of interesting objects to target.

Monthly Events

This section summarizes the events of each month, including Moon phases, conjunctions between the Moon and planets, conjunctions between planets and other planets, oppositions, eastern and western elongations, meteor shower peaks and much more.

All times are listed in UTC, or Coordinated Universal Time, which is the standard time used by astronomers throughout the year. The map shown below will guide you on how to calculate your local time from the time shown in UTC. It's important to note that most of North America will observe Daylight Saving Time (DST) between March 14, 2021, and November 7, 2021. Between these two dates, you will need to add an additional hour to your local time.

You'll notice some of the times listed fall during daylight, when observation is likely impossible. However, you can use this date

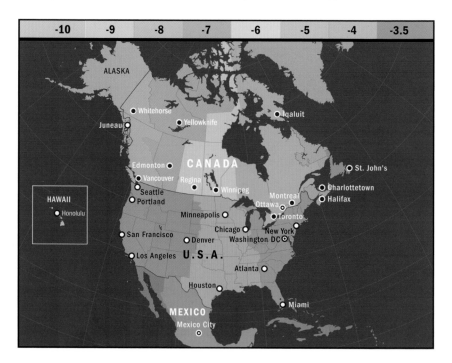

and time as a guide for observing an event on either the preceding night or the following night. Similarly, the coordinates given as the distance between celestial objects should be used as a guide and will not reflect exactly what you will see in the night sky. This is particularly true for close approaches that take place during the day, local time. For example, the calendar of events might say Mercury is 1.5 degrees south of Jupiter at 21:00 UTC, but by the time Mercury and Jupiter become visible to an observer in Eastern Time, that separation might have increased to 2.4 degrees.

This section also features a calendar layout with the Moon phases for each day shown, as well as a description of interesting Moon events for the month. You might notice some discrepancies between the coordinates given in these descriptions and those listed in the table of events. Because the Moon moves in right ascension so quickly, conjunctions between the Moon and planets as listed in the table may appear different for North American observers, and the separation between the objects may be more than suggested. The coordinates in the descriptions have been altered to more closely reflect the view from North America, but of course there may be more variation based on when the objects are visible to you in the night sky.

The Moon phases shown in the calendar are based on the day they occur in UTC time. For some observers in North America, that might mean the ideal time to watch, for example, the full Moon rising would be the night before.

Sky Charts and Descriptions

Each month features two sky charts, one facing south and the other facing north. These charts highlight select constellations, stars, clusters, nebulae, planets and galaxies visible in the night sky. While these sky charts are an accurate representation of the sky, it should be noted that factors like light pollution, smoke, cloud cover and so on can affect how much you see, and some of the fainter stars shown on the charts may not be visible.

The charts are drawn at a latitude of 45 degrees north. Observers north or south of this latitude will see slightly more of the northern or southern sky, respectively. The charts show the sky at 10:00 p.m. local standard time on the 15th of each month. You will also have the same view of the sky at 11:00 p.m. local standard time at the beginning of each month and at 9:00 p.m. local standard time at the end of each month. We note these times on the sky charts and also include Daylight Saving Time in parentheses between March and November. The planets shown on the sky charts will move slightly relative to the stars over the course of the month.

The preceding month's charts can be used for sky viewing two hours earlier, and the following month's charts can be used for sky viewing two hours later. So, for example, if you wanted to view the sky at 8:00 p.m. local time in February, you would refer to January's sky chart. If you're referring to a previous or later month's sky chart, note that the positions of the planets will not be correct.

January

January's Events

For those in the Northern Hemisphere, the winter months — though chilly in some parts — provide the best opportunity to observe the cosmos. Not only are the nights longer and the air clearer, but there are some magnificent targets, including the spectacular Quadrantid meteor shower peak on the night of January 2 to 3. The only challenge for some will be the clouds.

Calendar of Events

Day	Time (UTC)	Event
2	08:59	Earth at perihelion: 0.9833 astronomical units (au)
2–3		Quadrantid meteor shower peak
6	09:37	Last quarter of the Moon
9	15:39	Moon at perigee: 367,387 km (228,284 mi.)
9	21:00	Mercury 1.7°S of Saturn
11	11:00	Mercury 1.5°S of Jupiter
11	20:11	Venus 1.5°N of Moon
13	05:00	New Moon
20	21:02	First quarter of the Moon
21	13:11	Moon at apogee: 404,360 km (251,258 mi.)
24	01:59	Mercury 18.6°E of Sun (greatest eastern elongation)
24	02:26	Saturn in conjunction with Sun
28	19:16	Full Moon
29	00:51	Jupiter in conjunction with Sun

The Quadrantid meteor shower

The Moon This Month

SUN	MON	TUES	WED	THURS	FRI	SAT
					1	2
3	4	5	6 3rd Quarter	7	8	9
10	11	12	13 New Moon	14	15	16
17	18	19	20 1st Quarter	21	22	23
24	25	26	27	28 Full Moon	29	30
31						

We start off the month with an almost-full Moon (90 percent illuminated).

If you're an early riser, on January 11, you can catch a conjunction of Venus about 4 degrees east of a thin waning crescent Moon – which will only be roughly 4 percent illuminated. You can see the two celestial bodies in the early morning sky, low on the horizon. On the evening of January 14, the young crescent Moon low in the southwest can guide you to Mercury, which is about 7 degrees closer to the Sun, near the horizon. If you feel lucky, carry on and see if you can spot Jupiter and Saturn, which will soon disappear behind the Sun. This is the start of a favorable elongation of Mercury that unfolds through this month and into February. Mercury's greatest eastern elongation is on January 24.

The Moon will be at perigee (the closest the Moon will be in its monthly orbit) on January 9 and at apogee (the farthest it will be during its monthly orbit) on January 21.

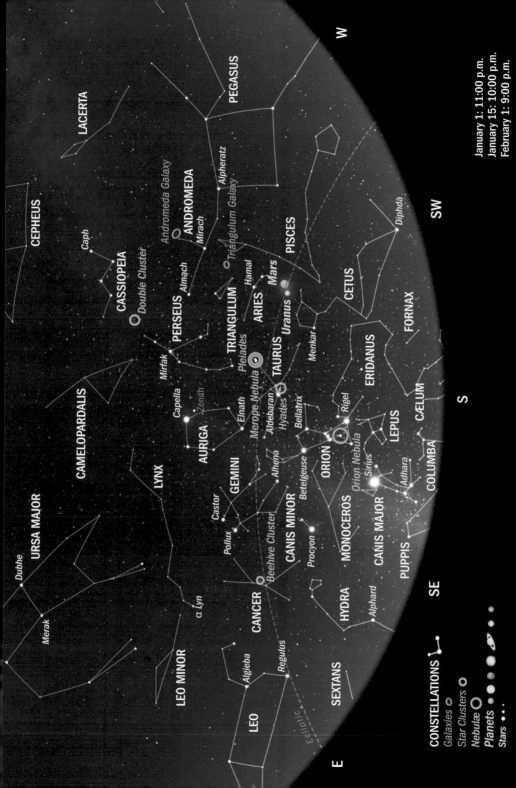

Highlights in the Southern Sky

This is a great month to check out **Mercury** as it reaches its greatest eastern elongation from the Sun on the night of January 23 to 24 at a magnitude of –0.6. Mercury will be at its highest point above the horizon in the southwest, on the evening of January 23.

Another key object this month is **Mars**, which lies high in the south-western sky in the constellation **Aries (the Ram)**. The fourth planet from the Sun is difficult to miss, as it shines fairly brightly, appearing as a red "star." During January, Mars will be at a magnitude of roughly –0.2, but as the month progresses, it will slowly begin to dim to a magnitude of around +0.4.

Orion (the Hunter) dominates the southern sky. The constellation is a veritable treasure trove of objects to observe, including the **Orion Nebula (Messier 42)**. The nebula is the brightest in the Northern Hemisphere. It appears as a bright smudge with binoculars, but it is far more defined using a telescope. It's even visible from light-polluted skies.

Orion's stars make the constellation easy to spot. The most famous star is **Betelgeuse**, a red supergiant lying 425 to 650 light-years away. The star, which is on the left "shoulder" of Orion, is 1,400 times larger and 4,000 times more luminous than our Sun. If Betelgeuse were at the center of our Solar System, the gas and dust that surrounds it would reach all the way to Neptune. Betelgeuse is an irregular long-period variable star that made the news in early 2020 for temporarily dimming by more than one magnitude in brightness.

The three stars in a line at the center of Orion — Alnitak, Alnilam and Mintaka (east to west) — make up the **Belt of Orion**. The vertical line below the belt is Orion's sword, which is where you'll find the Orion Nebula (M42). **Rigel**, the bright white star that makes up the left "foot" of Orion,

The Pleiades (Messier 45) and the Merope Nebula (NGC 1435)

is a blue supergiant. Rigel has two other companion stars, though these are invisible to the unaided eye.

The **Hyades (Melotte 25)** is a spectacular open star cluster to the northwest of Orion surrounding the bright red star **Aldebaran**, the eye of **Taurus (the Bull)**. Best viewed through binoculars, the cluster is a collection of roughly 100 stars that lies about 150 light-years from Earth.

Another marvelous star cluster is the famous **Pleiades (Messier 45)**, also known as the Seven Sisters. Though not as beautiful through bin-oculars as the Hyades, the cluster is still a fantastic sight. The blue glow of the **Merope Nebula (NGC 1435)** surrounding the cluster makes this a favorite target for many astrophotographers.

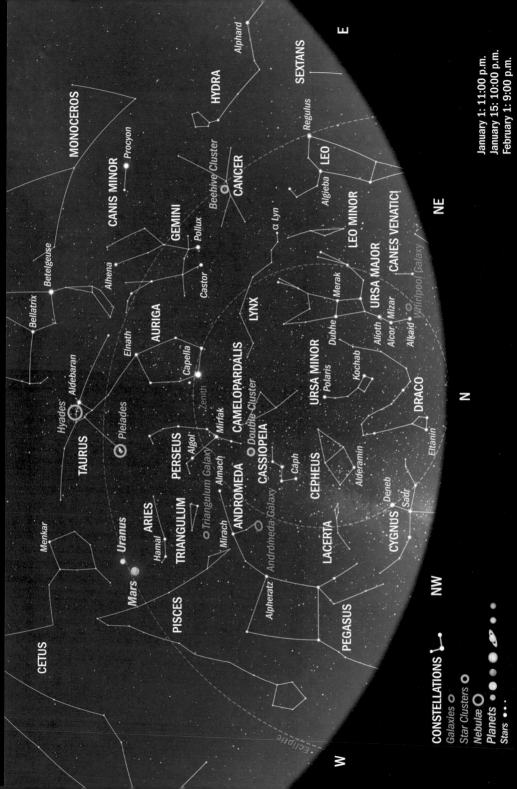

E

SEXTANS

HYDRA

Alphard

Procyon

CANIS MINOR

MONOCEROS

Betelgeuse

Bellatrix

CETUS

Menkar

CANCER

Beehive Cluster

GEMINI

Pollux

Castor

Alhena

AURIGA

Elnath

Capella

TAURUS

Hyades

Aldebaran

Pleiades

ARIES

Hamal

Uranus

Mars

PISCES

Regulus

LEO

Algieba

α Lyn

LYNX

Zenith

Mirfak

PERSEUS

Algol

Almach

Mirach

Triangulum Galaxy

TRIANGULUM

ANDROMEDA

Andromeda Galaxy

Alpheratz

PEGASUS

LEO MINOR

CANES VENATICI

Whirlpool Galaxy

URSA MAJOR

Merak

Dubhe

Alioth

Alcor Mizar

Alkaid

CAMELOPARDALIS

Double Cluster

CASSIOPEIA

Caph

CEPHEUS

Alderamin

LACERTA

CYGNUS

Deneb

Sadr

URSA MINOR

Polaris

Kochab

DRACO

Eltanin

NE

N

NW

W

CONSTELLATIONS
Galaxies ○
Star Clusters ○
Nebulæ ○
Planets ● ● ● ● ●
Stars ● • •

January 1: 11:00 p.m.
January 15: 10:00 p.m.
February 1: 9:00 p.m.

Ecliptic

The Whirlpool Galaxy (Messier 51)

Highlights in the Northern Sky

The **Quadrantid meteor shower** — which actually began December 27 — peaks on the night of January 2 to 3. The Quadrantids are one of the year's most active showers, with a ZHR of 120.

If you ever want to know which way is north, look for **Ursa Minor (the Little Bear)**, which is home to **Polaris**, better known as the "North Star." Polaris remains almost stationary in the sky as Earth rotates throughout the night, and you will notice the constellations revolving around the star month after month. Polaris is the star at the end of the handle of the **Little Dipper** asterism.

Ursa Major (the Great Bear), which contains the famous asterism called the **Big Dipper**, lies in the northeast. Though the Big Dipper appears to be a collection of seven stars, the second star in its handle contains two stars — **Alcor** and **Mizar** — that are very close together. Distinguishing these stars with the unaided eye was a traditional test of eyesight.

One of the most photographed galaxies in the region is the **Whirlpool Galaxy (Messier 51)**, which can be found near **Alkaid**, the star at the end of the Big Dipper's handle. This galaxy can be seen through a modest telescope, but see if you can detect it through binoculars.

Cassiopeia is an unmistakable constellation high in the northwest this month, making up a "W" shape. Not far from Cassiopeia is the **Andromeda Galaxy (Messier 31)**. Andromeda is one of the closest galaxies to our own Milky Way. It is easily visible without the aid of a telescope in dark-sky locations; the galaxy will look like a faint smudge in the sky. Andromeda has two satellite galaxies, **Messier 32** and **Messier 110**.

February

February's Events

The constellation Orion

February is another challenging month for some stargazers because of the cold and clouds. There are still a lot of things to enjoy during clear nights, however, including the constellations Orion, Gemini and Auriga. Though you may have to brave chillier temperatures, the good news is that there should be very little atmospheric disturbance due to those crisper nights.

This month, challenge yourself to observe northern-sky targets, particularly if you have a telescope. There are many hidden gems to be uncovered.

Calendar of Events

Day	Time (UTC)	Event
3	19:33	Moon at perigee: 370,116 km (229,979 mi.)
4	17:37	Last quarter of the Moon
8	13:39	Mercury at inferior conjunction
11	19:06	New Moon
18	10:22	Moon at apogee: 404,467 km (251,324 mi.)
18	22:47	Mars 4.1°N of Moon
19	18:47	First quarter of the Moon
23	7:38	Saturn 4°N of Mercury
27	8:17	Full Moon

The Moon This Month

SUN	MON	TUES	WED	THURS	FRI	SAT
	1	2	3	4	5	6
				3rd Quarter		
7	8	9	10	11	12	13
				New Moon		
14	15	16	17	18	19	20
					1st Quarter	
21	22	23	24	25	26	27
						Full Moon
28						

The Moon is at perigee on February 3, and the last quarter is the next day. On February 11, we have a new Moon, and just a week later, the Moon is at apogee. Try looking for a young crescent Moon low in the west after sunset on February 12.

At the beginning of February, Mercury may still be visible low in the west just after sunset, but it is quickly heading toward inferior conjunction with the Sun on February 8. A special treat occurs on the night of apogee – February 18 – with a conjunction of Mars and a nearly first-quarter Moon. The pair will be about 4 degrees apart that evening, and the open star clusters of the Hyades and the Pleiades will be sitting nearby.

The first quarter occurs on February 19 and the full Moon happens on February 27.

The Hyades (Melotte 25) star cluster

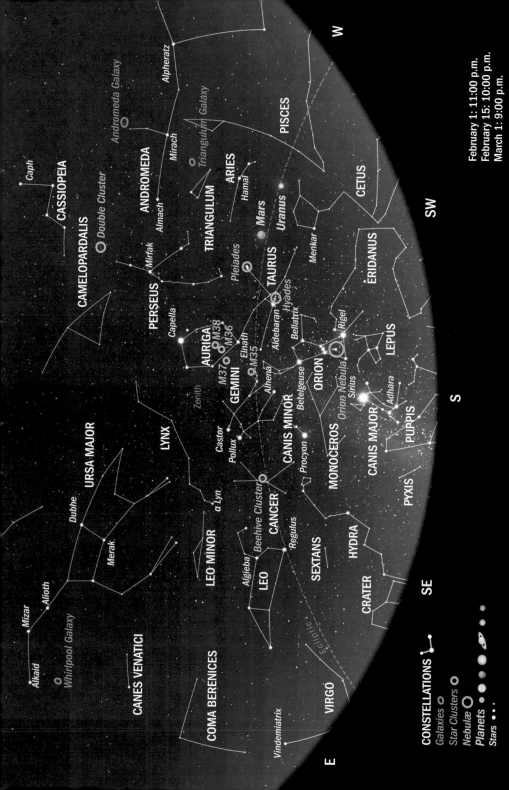

W

February 1: 11:00 p.m.
February 15: 10:00 p.m.
March 1: 9:00 p.m.

SW

S

SE

E

CONSTELLATIONS ⌐•
Galaxies ⊙
Star Clusters ⊙
Nebulae ⊙
Planets ●●●
Stars • • •

Caph
CASSIOPEIA
CAMELOPARDALIS
Double Cluster
Andromeda Galaxy
Alpheratz
Mirach
ANDROMEDA
Almach
Triangulum Galaxy
TRIANGULUM
ARIES
Hamal
Mirfak
PERSEUS
Pleiades
Mars
Uranus
TAURUS
PISCES
Capella
AURIGA
M38
M36
Elnath
Aldebaran
Hyades
Bellatrix
Menkar
CETUS
ERIDANUS
M37
M35
Athena
Rigel
Zenith
GEMINI
Betelgeuse
ORION
LEPUS
Castor
Pollux
CANIS MINOR
Orion Nebula
Sirius
Adhara
α Lyn
LYNX
Procyon
MONOCEROS
CANIS MAJOR
PUPPIS
URSA MAJOR
Dubhe
Beehive Cluster
CANCER
Regulus
SEXTANS
PYXIS
Merak
LEO MINOR
Algieba
LEO
HYDRA
Mizar
Alioth
Whirlpool Galaxy
Alkaid
CANES VENATICI
COMA BERENICES
Vindemiatrix
VIRGO
Ecliptic
CRATER

Highlights in the Southern Sky

Canis Major (the Great Dog) takes the spotlight in the south, with **Sirius** — the brightest star in the night sky — taking center stage. Like many stars, Sirius is part of a binary star system. There is Sirius A (which is the one we can see) and a much smaller white dwarf star, Sirius B, that is hidden to the naked eye due to the brightness of Sirius A. Known as the Dog Star, Sirius A is a blue-white star that lies 8.6 light-years from Earth and is roughly 25 times more luminous than our Sun.

Looking northward is the constellation **Monoceros**, a somewhat faint constellation. North of that lies **Canis Minor (the Little Dog)**, which contains another binary star system called **Procyon**.

Farther north lies the constellation **Gemini (the Twins)**. The two brightest stars in this constellation are **Castor** and **Pollux**. Castor is the bluer of the two stars and lies closer to the celestial pole. Another memory aid is that Castor is closer to Capella in Auriga, and Pollux is closer to Procyon. Castor is a fascinating system: rather than being a mere binary, it's actually a six-star system.

Look for the open cluster **Messier 35** with binoculars, near the foot of the twin Castor.

The constellation **Auriga (the Charioteer)** is prominent high in the southwest, with the bright star **Capella** gleaming as the sixth-brightest star in the night sky. This constellation straddles the winter Milky Way and contains many fine binocular views, including the three open clusters **Messier 36**, **Messier 37** and **Messier 38**.

A somewhat fainter **Mars** than we saw in January remains in the west.

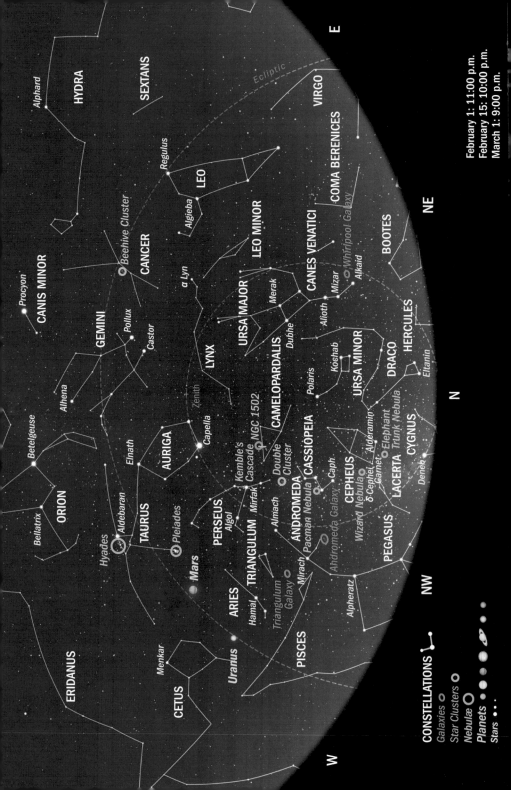

Highlights in the Northern Sky

The northern sky is often overlooked as a rich region of the Milky Way, but there are many offerings here.

Running through **Cassiopeia** down to the horizon is a region full of clusters and some beautiful nebulae. Examples of nebulae include the **Pacman Nebula (NGC 281)** located in Cassiopeia, as well as the **Wizard Nebula (NGC 7380)** and the **Elephant Trunk Nebula (IC 1396A)** located in **Cepheus**. These nebulae are favorite targets for experienced astrophotographers and are not easy objects to spot for beginners. There are also a few binocular objects in this region to enjoy, like **Mu Cephei** (the "Garnet Star"), the open cluster **Messier 103** as well as the variable double star **Delta Cephei**.

Perseus is high in the northwestern sky and offers up a fantastic binocular target, the **Double Cluster (NGC 884 and NGC 869)**. This pair of open star clusters can be found between Cassiopeia and Perseus and are marvelous to see through binoculars. If you don't have a pair of binoculars, try to head out to a dark-sky location, where the clusters can be visible to the unaided eye. Perseus is also home to **Algol**, the first-known eclipsing variable star, with a precise period of two days, 20 hours and 49 minutes. Winter is the perfect season to watch the fading and brightening of this star in a single night, and it is high up and easy to observe with the unaided eye.

In the northeast are the constellations **Canes Venatici (the Hunting Dogs)**, **Coma Berenices (Berenice's Hair)** and **Virgo (the Maiden)**. This region is rich with galaxies. The **Coma Cluster** contains thousands of galaxies and is found in Coma Berenices.

The obscure constellation **Camelopardalis (the Giraffe)** is home to **Kemble's Cascade**, an asterism of a string of stars ending in the open cluster **NGC 1502**. The asterism was named for Canadian amateur astronomer Father Lucien Kemble, who first noted it. This is an easy binocular target.

The Coma Cluster

March

March's Events

As March rolls in, the more familiar winter constellations are starting to make their way to the west, with springtime constellations — and if you're willing to brave early mornings, summer constellations — making their way into our night sky once again.

There's a lot of planet-gazing to do this month, as long as you're willing to get up early. Mercury, Saturn and Jupiter will all be visible in the eastern dawn sky. On March 9, the planets line up: Mercury, Jupiter and Saturn (in order) near a thin crescent Moon.

Mars is still a great sight in the northwest evening sky, in between the Hyades and the Pleiades star clusters. On March 19, it will lie roughly 3 degrees northwest of a thick crescent Moon.

The vernal (spring) equinox occurs on March 20.

Calendar of Events

Day	Time (UTC)	Event
2	05:19	Moon at perigee: 365,423 km (227,063 mi.)
3	23:32	Mars 2.6°S of Pleiades
5	05:11	Mercury 0.3°N of Jupiter
6	01:30	Last quarter of the Moon
6	10:59	Mercury 27.3°W of Sun (greatest western elongation)
9	23:02	Saturn 3.9°N of Moon
10	15:35	Jupiter 4.3°N of Moon
10	23:36	Neptune in conjunction with Sun
11	01:02	Mercury 3.9°N of Moon
13	10:21	New Moon
14		Daylight Saving Time begins (most of U.S. and Canada)
18	05:04	Moon at apogee: 405,252 km (251,812 mi.)
19	17:48	Mars 2.1°N of Moon
20	09:37	Vernal (spring) equinox
20	19:15	Mars 6.9°N of Aldebaran
21	14:40	First quarter of the Moon
26	06:17	Venus at superior conjunction
28	18:48	Full Moon
30	06:12	Moon at perigee: 360,309 km (223,886 mi.)

The Moon This Month

SUN	MON	TUES	WED	THURS	FRI	SAT
	1	2	3	4	5	6 *3rd Quarter*
7	8	9	10	11	12	13 *New Moon*
14	15	16	17	18	19	20
21 *1st Quarter*	22	23	24	25	26	27
28 *Full Moon*	29	30	31			

The month starts off with the Moon at last quarter on March 6. The new Moon is a week later on March 13. The Moon will be at perigee twice this month: on March 2 as well as on March 30. It reaches apogee on March 18.

Early on March 10, Jupiter will be roughly 4 degrees north of the waning crescent Moon, low in the southeastern dawn sky, making for a lovely pairing together with Mercury to the east and Saturn to the west. Then on March 19, Mars is about 3 degrees northwest of the Moon, with the Hyades nearby.

Seeing multiple planets pairing with the Moon is an enchanting experience, like this Moon-Venus-Mercury combo in the dawn sky.

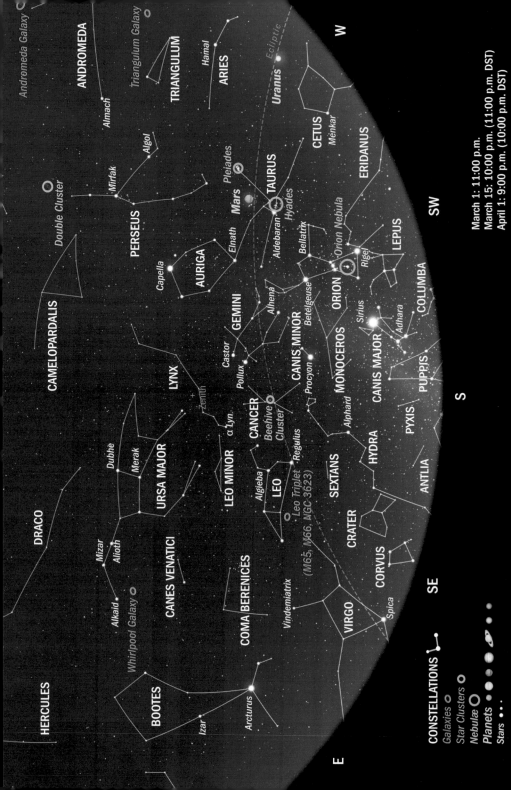

CONSTELLATIONS ⌐•⌐
Galaxies ⊙
Star Clusters ○
Nebulæ ◎
Planets ● ● ● ◆
Stars ● ● •

March 1: 11:00 p.m.
March 15: 10:00 p.m. (11:00 p.m. DST)
April 1: 9:00 p.m. (10:00 p.m. DST)

W

SW

S

SE

E

HERCULES

BOOTES

Izar

Arcturus

DRACO

CANES VENATICI

Alkaid

Mizar

Alioth

Whirlpool Galaxy ⊙

COMA BERENICES

Vindemiatrix

VIRGO

Spica

CORVUS

CRATER

URSA MAJOR

Dubhe

Merak

LEO MINOR

Algieba

LEO

Regulus

Leo Triplet

(M65, M66, NGC 3623)

SEXTANS

HYDRA

Alphard

ANTLIA

PYXIS

LYNX

α Lyn.

Zenith

CANCER

Beehive Cluster ⊙

Castor

Pollux

GEMINI

Alhena

CAMELOPARDALIS

Double Cluster ⊙

PERSEUS

Mirfak

Algol

AURIGA

Capella

Elnath

TAURUS

Aldebaran

Hyades

Pleiades

Mars

CANIS MINOR

Procyon

MONOCEROS

Betelgeuse

ORION

Bellatrix

Orion Nebula ◎

Rigel

Sirius

CANIS MAJOR

Adhara

PUPPIS

COLUMBA

LEPUS

ERIDANUS

CETUS

Menkar

Uranus

Ecliptic

ANDROMEDA

Andromeda Galaxy ⊙

TRIANGULUM

Triangulum Galaxy ⊙

Almach

ARIES

Hamal

The Beehive Cluster (Messier 44)

Highlights in the Southern Sky

This month, **Hydra (the Water Snake)**, the longest of the 88 constellations, lies to the southeast. The **Beehive Cluster (Messier 44)**, also known as Praesepe (the Manger), lies in the constellation **Cancer (the Crab)**. Containing roughly 1,000 stars, this open star cluster is just 600 light-years away, making it one of the closest to Earth.

Leo (the Lion) is high in the southeast, east of Hydra, with its bright star **Regulus** shining brightly. Leo is home to many galaxies, including a collection of three galaxies called the **Leo Triplet: Messier 65, Messier 66** and **NGC 3623**. Again, a modest telescope can reveal the trio.

Virgo (the Maiden), the second-largest constellation in the night sky with its bright star **Spica**, lies low in the east and begins its march higher in the sky this month. Interestingly, in 1992, the very first exoplanets (planets outside our Solar System) were discovered around a pulsar designated PSR B1257+12, in the direction of Virgo. Since then, many more exoplanets have been found hiding in the constellation.

High in the south is **Gemini (the Twins)**, with **Castor** and **Pollux**. To the west is the constellation **Auriga (the Charioteer)**, containing the bright star **Capella**.

Orion (the Hunter) is beginning to make its way to the southwest, along with the **Hyades (Melotte 25)** and the **Pleiades (Messier 45)**, which lie to the west. One big treat this month is the position of **Mars**, which will sit between the Hyades and the Pleiades.

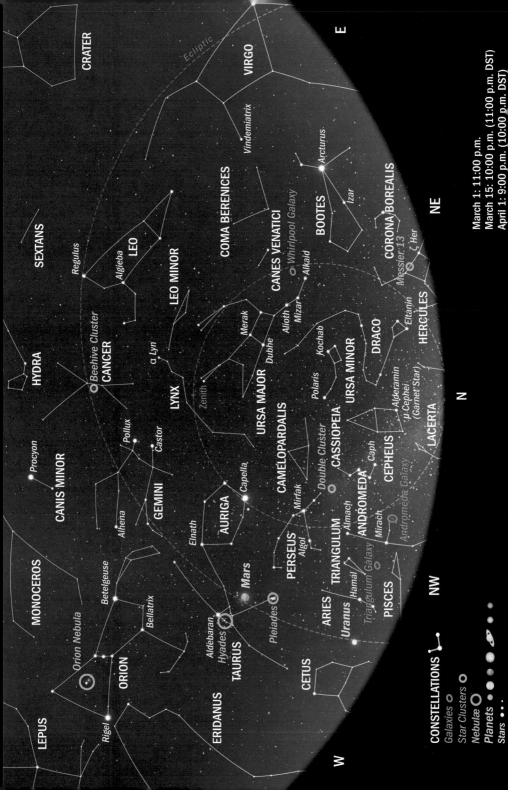

CRATER

Ecliptic

VIRGO

E

SEXTANS

Vindemiatrix

Arcturus.

LEO
Regulus
Algieba

COMA BERENICES

CANES VENATICI

BOOTES

Izar

NE

CORONA BOREALIS

LEO MINOR

Whirlpool Galaxy
Alkaid

Messier 13
ζ Her

HYDRA

Beehive Cluster
CANCER

α Lyn

Merak
Dubhe

Alioth
Mizar

HERCULES

Eltanin

LYNX

Zenith

URSA MAJOR

Kochab

DRACO

Pollux

CAMELOPARDALIS

Polaris

URSA MINOR

Alderamin
μ Cephei
(Garnet Star)

N

Procyon

Castor

Double Cluster

CASSIOPEIA

LACERTA

CANIS MINOR

GEMINI

Capella

Caph

CEPHEUS

Alhena

AURIGA

Mirfak

ANDROMEDA

Andromeda Galaxy

Elnath

PERSEUS

Almach

Betelgeuse

Algol

TRIANGULUM

Mirach

MONOCEROS

Mars

Triangulum Galaxy

PISCES

Bellatrix

ARIES
Hamal

Uranus

NW

Pleiades

Orion Nebula

Aldebaran
Hyades

ORION

TAURUS

CETUS

ERIDANUS

LEPUS

Rigel

W

CONSTELLATIONS
Galaxies ○
Star Clusters ○
Nebulæ ○
Planets ● ● ● ● ●
Stars ● ● ·

March 1: 11:00 p.m.
March 15: 10:00 p.m. (11:00 p.m. DST)
April 1: 9:00 p.m. (10:00 p.m. DST)

Highlights in the Northern Sky

The constellation **Boötes (the Herdsman)**, which is pronounced boo-oh-tees, is low on the eastern horizon this month. Boötes is a large constellation that is home to the fourth-brightest star in the night sky, **Arcturus**. You can use the **Big Dipper** to find it by just following the curve of the handle — the stars **Alioth**, **Mizar** and **Alkaid**.

Arcturus, which means "guardian of the bear," is an orange giant star and a double star that can be seen through a telescope. The larger of the double stars is 25 times larger than our Sun and is believed to be roughly 7.1 billion years old. Arcturus is also an interesting system, as it moves at a different angle and speed than most of the stars in our galaxy. Arcturus is believed to be part of a group of several stars that move at a similar angle and speed named the Arcturus Moving Group, which is thought to have been created when a dwarf galaxy collided with the Milky Way.

The constellation **Cepheus** can be found directly north this month, to the right of **Cassiopeia**.

The star Arcturus

March 2021

April

April's Events

April is a great time to continue your planetary viewing. If you're willing to get up in the early morning, Saturn and Jupiter both have conjunctions with the Moon. Mars makes a close conjunction with the Moon on April 17 (though not seen from North America), and the Moon even covers Mars as seen from parts of Africa and Southeast Asia.

The night of April 21 to 22 is the peak of the Lyrid meteor shower. Though it's not a particularly strong shower — with a ZHR of roughly 18 — it can produce some fireballs.

Calendar of Events

Day	Time (UTC)	Event
4	10:02	Last quarter of the Moon
6	08:34	Saturn 4.2°N of Moon
7	07:15	Jupiter 4.7°N of Moon
12	02:31	New Moon
14	17:47	Moon at apogee: 406,119 km (258,564 mi.)
17	12:09	Mars 0.1°N of Moon
19	01:31	Mercury at superior conjunction
20	06:59	First quarter of the Moon
21–22		Lyrid meteor shower peak
27	03:31	Full Moon
27	15:24	Moon at perigee: 357,378 km (222,064 mi.)
30	21:15	Uranus in conjunction with Sun

The constellation Leo dominates the southern sky in April and is a great time to view its trio of galaxies, Messier 65, Messier 66 and NGC 3623.

The Moon This Month

SUN	MON	TUES	WED	THURS	FRI	SAT
				1	2	3
4 3rd Quarter	5	6	7	8	9	10
11	12 New Moon	13	14	15	16	17
18	19	20 1st Quarter	21	22	23	24
25	26	27 Full Moon	28	29	30	

The month begins with the last quarter on April 4. But a couple days later, on April 6, Saturn will be roughly 5 degrees north of the crescent Moon. The next day, it's Jupiter's turn for a close encounter with the Moon. The pair will be separated by about 5 degrees.

The new Moon occurs five days later on April 12, the first quarter on April 20 and the full Moon on April 27. Apogee occurs on April 14, and perigee occurs on April 27, the day of the full Moon. Seaside residents may see large ocean tides in the following days.

The full Moon above Pigeon Point Lighthouse in California

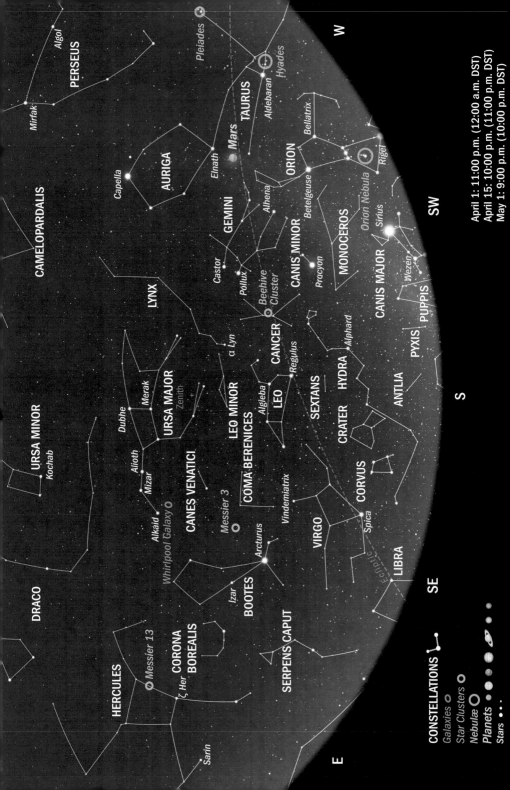

Messier 3

Highlights in the Southern Sky

The constellation **Boötes (the Herdsman)** is now clearly visible higher in the east by April, as is **Virgo (the Maiden)**. The brightest star in Virgo is **Spica**, which is roughly 10 times more massive than the Sun and lies 260 light-years from us. Spica is another example of a binary star, where two stars orbit around a shared point. But the second, smaller star isn't visible — not even to a large telescope. The pair are extremely close together, separated by roughly 17 million kilometers (10.5 million miles). That's just a third of the average distance between Mercury and the Sun.

Coma Berenices (Berenice's Hair) sits to the right of Boötes, but a special treat lies in between the two constellations: a spectacular globular cluster called **Messier 3**, the first object to be discovered by Charles Messier, the astronomer who compiled the Messier Catalog of objects. At first, Messier mistook the globular cluster of stars for a nebula, but that was corrected by astronomer William Herschel 20 years later, in 1784. It's believed that Messier 3 contains more than 500,000 stars. It also contains more variable stars — stars that change brightness over time — than any other cluster.

The constellation **Corona Borealis (the Northern Crown)**, lies to the east of Boötes. Looking like a cup, its brightest star is **Alphecca**, a binary. The star is part of an eclipsing binary system — meaning every so often, the second star passes in front of Alphecca, causing it to dim slightly. There is also evidence that there is a disk around the stars that could potentially be forming planets.

The constellation **Leo (the Lion)** dominates the southern sky, with **Gemini**'s twin stars **Castor** and **Pollux** now part of the western sky.

The **Beehive Cluster (Messier 44)** is in a great position in **Cancer (the Crab)** for binocular viewing. **Mars** is also still visible high in the west.

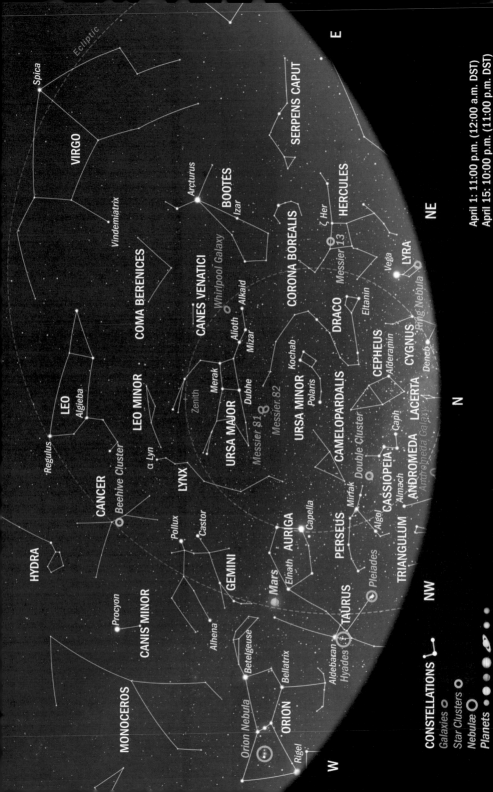

Messier 81 and Messier 82

Highlights in the Northern Sky

Perseus is low in the northwestern sky, together with another constellation — **Auriga (the Charioteer)**. However, Auriga's brightest star, **Capella**, still lights up the sky.

Cassiopeia is also beginning to sink lower in the north, while **Cepheus** still dominates the north.

Ursa Major (the Great Bear) and its asterism, the **Big Dipper**, are now upside-down high in the north. **Draco (the Dragon)** weaves between Ursa Major and **Ursa Minor (the Little Bear)**; **Polaris** is visible as the North Star in the latter constellation.

There are a few galaxies in the northern sky that are great targets for telescopes or astrophotographers, particularly during April. **Messier 81 (Bode's Galaxy)**, found in Ursa Major, is one of the brightest in the night sky. M81 can be seen from dark-sky sites through binoculars or a small telescope. **Messier 82**, or the Cigar Galaxy, will be in the same field of view.

Hercules rises in the northeast this month. One of the most beautiful globular clusters is **Messier 13**, which lies in this constellation. The cluster is visible as a faint fuzz in dark-sky locations, but you can also use binoculars. Even a small telescope will reveal the tight ball of stars.

May

May's Events

The shorter nights might mean less time for astronomical viewing, but this is a good time to start planning out your summer targets.

There is a meteor shower, the Eta Aquariids, which began on April 19 but will peak on the night of May 4 to 5. The shower could produce 10 to 30 meteors an hour, though it's better seen from the Southern Hemisphere.

On May 13, you might try to catch Mercury. It will be about 3 degrees right of the Moon, low in the western evening sky, when it is dark enough to see. The Moon, however, will be less than 1 percent illuminated, so finding the planet could be a challenge. Mercury is best observed around the date of its greatest eastern elongation on May 17. If you want a greater challenge, try to look for the conjunction of Venus and the Moon (just over 24 hours old), low in the west on the evening of May 12, just after sunset.

On May 26, there is a total lunar eclipse visible across parts of North America, though only in partiality as the Moon sets. To learn more about this event, refer to page 27.

On the evening of May 28, look for a lovely conjunction of Venus and Mercury low in the northwest just after sunset.

Calendar of Events

Day	Time (UTC)	Event
3	17:02	Saturn 4.4°N of Moon
3	19:50	Last quarter of the Moon
4	21:00	Jupiter 4.9°N of Moon
5–6		Eta Aquariid meteor shower peak
11	19:00	New Moon
11	21:54	Moon at apogee: 406,512 km (252,595 mi.)
13	17:59	Mercury 2.4°N of Moon
16	04:47	Mars 1.6°S of Moon
17	05:59	Mercury 22°E of Sun (greatest eastern elongation)
19	19:13	First quarter of the Moon
26	01:52	Moon at perigee: 357,311 km (222,023 mi.)
26	11:14	Full Moon
26	11:18	Total lunar eclipse
29	03:01	Venus 0.4°N of Mercury
30	01:22	Saturn 4.3°N of Moon

The Moon This Month

SUN	MON	TUES	WED	THURS	FRI	SAT
						1
2	3 3rd Quarter	4	5	6	7	8
9	10	11 New Moon	12	13	14	15
16	17	18	19 1st Quarter	20	21	22
23	24	25	26 Full Moon	27	28	29
30	31					

This month we continue our planetary viewing with the giants of the Solar System, Saturn and Jupiter, which are dancing with the Moon. On May 3, Saturn is roughly 6 degrees north of the Moon, which is moving out of its last quarter. The next day, the Moon lies in between the two planets, though slightly to the south. Two days later, on May 6, Jupiter is about 8 degrees north of the Moon just before sunrise.

On the evening of May 15, it's Mars that has a particularly close encounter — less than 2 degrees — with the Moon. Then on May 31, Saturn and the Moon have yet another close encounter in the early morning sky.

A new Moon occurs on May 11, and it's also at apogee. The Moon's first quarter occurs on May 19, with the full Moon occurring on May 26; that day is also lunar perigee, so the Moon with appear at its largest in 2021, and seaside residents may see large ocean tides in the following days.

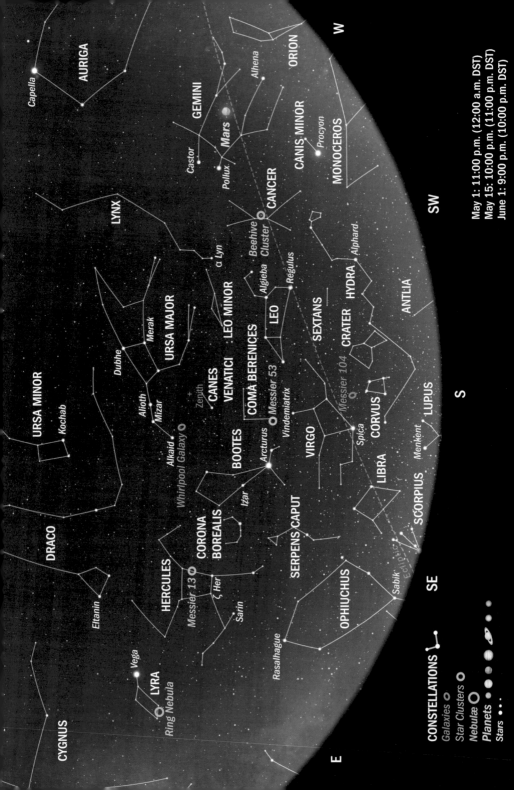

W

AURIGA
Capella

ORION
Alhena

GEMINI

Castor
Mars
Pollux
Procyon

CANIS MINOR

LYNX
α Lyn

MONOCEROS

Beehive
Cluster

CANCER

SW

URSA MINOR
Kochab

URSA MAJOR
Dubhe
Merak
Alioth
Mizar
Alkaid
Whirlpool Galaxy
Zenith

CANES
VENATICI

LEO MINOR

Algieba

LEO

Regulus

COMA BERENICES
Messier 53
Vindemiatrix

SEXTANS

HYDRA
Alphard

DRACO
Ettanin

BOOTES
Izar
Arcturus

VIRGO
Spica

Messier 104
CRATER

ANTLIA

S

CORONA
BOREALIS

HERCULES
Messier 13
ζ Her
Sarin

SERPENS CAPUT

CORVUS

LIBRA

Menkent
LUPUS

Vega
LYRA
Ring Nebula

CYGNUS

OPHIUCHUS
Rasalhague

Sabik
Ecliptic

SCORPIUS

E

SE

CONSTELLATIONS
Galaxies ○
Star Clusters ○
Nebulæ ○
Planets ● ● ● ●
Stars ● ● ●

May 1: 11:00 p.m. (12:00 a.m. DST)
May 15: 10:00 p.m. (11:00 p.m. DST)
June 1: 9:00 p.m. (10:00 p.m. DST)

Messier 53

Highlights in the Southern Sky

It's Virgo's **(the Maiden)** time to shine, with its bright star **Spica** clearly visible high in the southern sky. However, **Boötes's (the Herdsman)** star **Arcturus** remains a brighter star in the sky, hovering east of Virgo. Above Virgo is **Coma Berenices (Berenice's Hair)**. Within this constellation lies **Messier 53**, a tight globular cluster that has the distinction of being one of the most distant globular clusters, at 59,700 light-years away. The cluster can be seen through small telescopes, though larger telescopes will reveal the individual stars with more clarity.

The constellation **Hydra (the Water Snake)** is now low on the southern horizon, slinking along to the west. Above Hydra are the small constellations **Corvus (the Crow)** and **Crater (the Cup)**. Straddling Corvus and Virgo is the spectacular **Sombrero Galaxy (Messier 104)**. This galaxy is visible through a small telescope edge-on, with an incredible view of its dust lane best viewable through larger telescopes. The Sombrero Galaxy is 28 million light-years away and has the mass of 800 billion Suns, making it one of the most massive objects in the Virgo galaxy cluster.

Along with **Ophiuchus (the Serpent Bearer)**, the constellation **Libra (the Scales)** is low in the southeast. Though Libra is a faint constellation, it sits above **Scorpius (the Scorpion)**, which will soon rise higher in the south with its load of treasures.

The twin stars **Castor** and **Pollux** are now fixtures in the western sky.

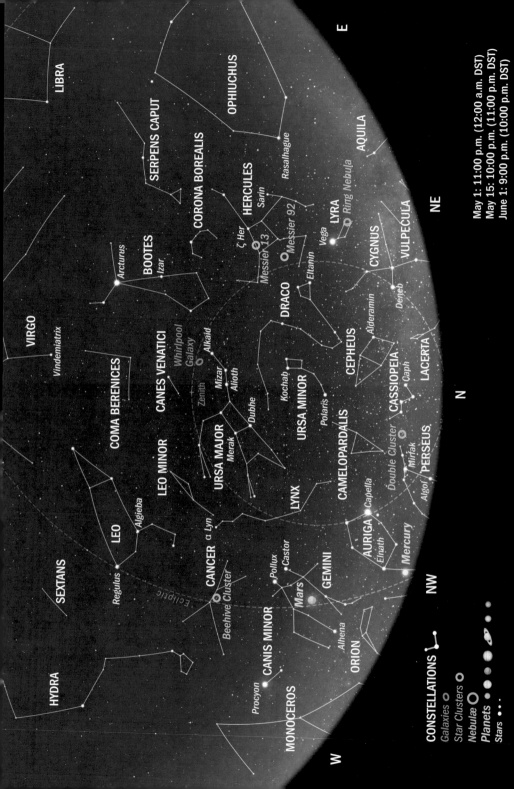

The Ring Nebula (Messier 57)

Highlights in the Northern Sky

The constellation **Hercules** gives you a fantastic opportunity to observe the globular cluster **Messier 13** with a pair of binoculars or, if you're at a dark-sky location, with just the naked eye. But there's another nearby bright globular cluster, called **Messier 92**. That cluster contains roughly 300,000 tightly packed stars. M92 lies to the west of Hercules and the constellation's four-star asterism, the **Keystone.**

Directly east of M92 is the constellation **Lyra (the Lyre)**, with **Vega**, the fifth-brightest star in the sky, shining brightly. In popular culture, Vega was featured as the location of an alien signal in Carl Sagan's book *Contact* (as well as the movie of the same name).

Lyra also is home to the **Ring Nebula (Messier 57)**. This planetary nebula looks like what its name suggests, and it lies about 2,000 light-years away. Though not visible to the unaided eye, the Ring Nebula is visible in small telescopes.

Cygnus (the Swan) is higher in the northeast with its bright star **Deneb**, which represents the tail of the swan. The star, roughly 200 times more massive than our Sun, is one of the most distant stars we can see with the naked eye, lying some 1,500 light-years away.

Cepheus has now moved farther to the northeast, while **Cassiopeia** lies low in the north and **Perseus** sinks lower in the northwest. However, the **Double Cluster (NGC 869 and NGC 884)** still remains above the horizon between Perseus and Cassiopeia.

The constellation **Auriga (the Charioteer)** and its bright star **Capella** are low in the northwest and will eventually disappear from view.

June

June's Events

Summer is upon us, and so are numerous targets of the summer sky. The planets are first, with viewings of Jupiter, Saturn and Mars throughout the month.

But more exciting is the rising of two constellations Sagittarius and Scorpius in the southern sky. The two constellations mark some of the richest areas for any type of viewing, whether it be unaided or with the help of binoculars or a telescope.

On June 10 northern parts of North America will be treated to an annular solar eclipse. To learn more about this event, refer to page 29.

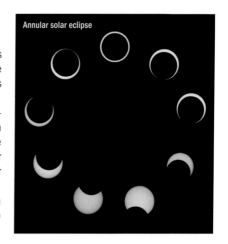

Annular solar eclipse

Calendar of Events

Day	Time (UTC)	Event
1	08:57	Jupiter 4.9°N of Moon
2	07:24	Last quarter of the Moon
8	02:27	Moon at apogee: 406,228 km (252,418 mi.)
10		Annular solar eclipse
10	10:53	New Moon
11	01:06	Mercury at inferior conjunction
12	06:44	Venus 1.6°S of Moon
13	19:52	Mars 3°S of Moon
18	03:54	First quarter of the Moon
21	03:32	Summer solstice
23	09:58	Moon at perigee: 359,956 km (223,666 mi.)
24	18:40	Full Moon
27	09:30	Saturn 4.1°N of Moon
28	18:38	Jupiter 4.6°N of Moon

The Moon This Month

SUN	MON	TUES	WED	THURS	FRI	SAT
		1	2	3	4	5
				3rd Quarter		
6	7	8	9	10	11	12
				New Moon		
13	14	15	16	17	18	19
					1st Quarter	
20	21	22	23	24	25	26
				Full Moon		
27	28	29	30			

At this time of year, the planets are shifting times. Instead of appearing only in the early morning, they rise after midnight and are still visible high in the southwest before sunrise. Such is the case on June 1, when the almost-last-quarter Moon and Jupiter are around 6 degrees apart.

On June 13, Mars will be roughly 3 degrees south of the waxing crescent Moon and low in the western evening sky.

The Moon will be at apogee on June 8, just before it becomes new. The new Moon occurs on June 10, with the first quarter on June 18. The full Moon occurs on June 24, just after it has reached perigee.

The constellations Scorpius and Sagittarius begin to rise in the June sky

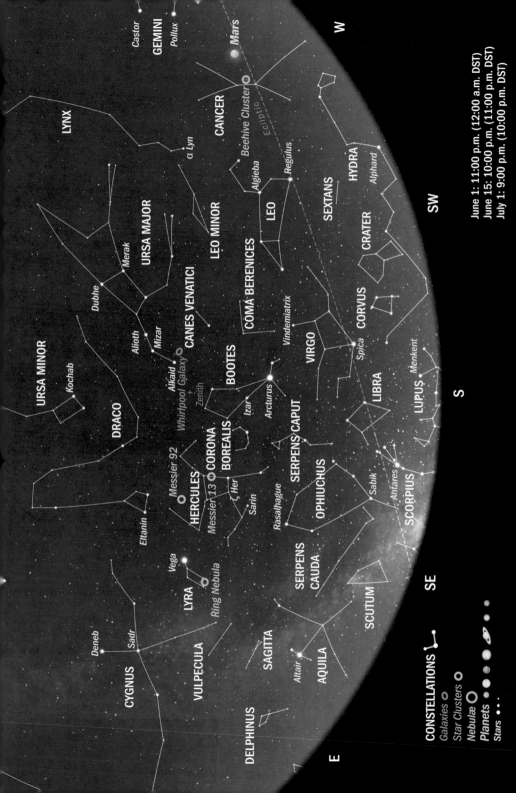

The center of the Milky Way with bright red Antares in the constellation Scorpius

Highlights in the Southern Sky

With **Libra (the Scales)** now higher in the southern sky, you might notice a red star twinkling slightly to the east of it, near the horizon. It's not Mars, but rather **Antares**, or the "rival of Mars." This star is the heart of **Scorpius (the Scorpion)**, which is a marvelous summertime constellation rich with naked-eye, binocular and telescopic targets. Antares is a massive star — a red supergiant like Orion's Betelgeuse — and is several hundred times the diameter of the Sun, along with being roughly 10,000 times more luminous. Lying some 600 light-years away, Antares is nearing the end of its life. As a red supergiant star, it will eventually die in a spectacular explosion called a supernova. Antares has a smaller, dimmer companion called Antares B.

The constellation **Ophiuchus (the Serpent Bearer)** is now higher in the southeast. **Boötes (the Herdsman)** is also high in the southeast, and its star **Arcturus** is the brightest star in the southern sky.

Hercules has now moved off to the east, which provides a great opportunity to take in the globular clusters **Messier 13** and **Messier 92**.

The constellation **Hydra (the Water Serpent)** is very near the southern horizon. **Leo (the Lion)** is setting in the west together with **Cancer (the Crab)**, as the constellations get ready to dip below the horizon.

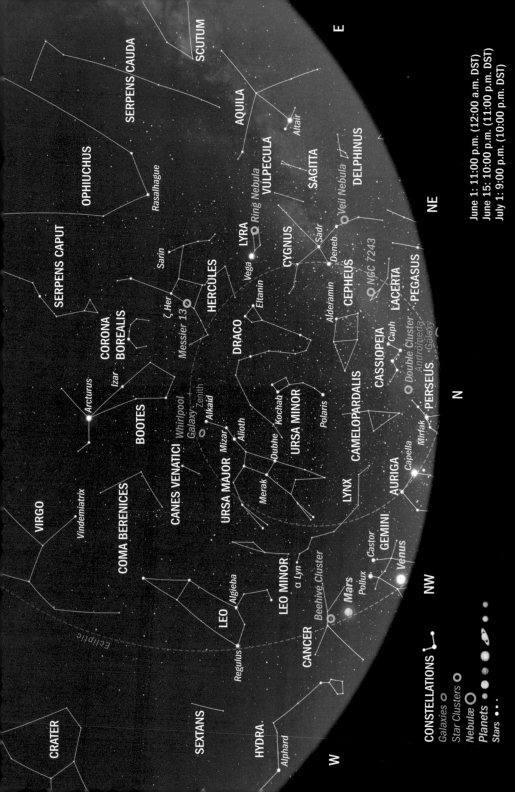

E

SCUTUM

SERPENS CAUDA

AQUILA

Altair

OPHIUCHUS

VULPECULA

SAGITTA

DELPHINUS

Rasalhague

Ring Nebula

NE

LYRA

CYGNUS

Sadr

CEPHEUS

SERPENS CAPUT

Sarin

Vega

Deneb.

NGC 7243

LACERTA

Eltanin

Aldemain

PEGASUS

CORONA
BOREALIS

ζ. Her

HERCULES

Messier 13

DRACO

Caph

Double Cluster

Andromeda

Izar

Galaxy

CASSIOPEIA

BOOTES

Zenith

PERSEUS

Arcturus

Whirlpool

Alkaid

Kochab

Mirfak

Galaxy

Dubhe

N

VIRGO

CANES VENATICI

Mizar

URSA MINOR

CAMELOPARDALIS

Capella

Alioth

Polaris

Vindemiatrix

URSA MAJOR

AURIGA

COMA BERENICES

Merak

LYNX

Algieba

LEO MINOR

GEMINI

Venus

LEO

α Lyn

Castor

NW

Beehive Cluster

Pollux

Mars

Regulus

CANCER

SEXTANS

HYDRA.

CRATER

Alphard

Ecliptic

W

June 1: 11:00 p.m. (12:00 a.m. DST)
June 15: 10:00 p.m. (11:00 p.m. DST)
July 1: 9:00 p.m. (10:00 p.m. DST)

CONSTELLATIONS

Galaxies ○
Star Clusters ○
Nebulæ ○
Planets ●
Stars ·

Noctilucent clouds

Highlights in the Northern Sky

The constellation **Ursa Major (the Great Bear)** now hangs sideways in the northwest. The **Whirlpool Galaxy (Messier 51a)**, which is found near the star **Alkaid**, is a great target for telescopes given the galaxy's position so high in the north.

The constellation **Aquila (the Eagle)** is rising in the northeast with its bright star **Altair**, which means all three stars that form the **Summer Triangle** are visible. Altair, together with **Vega** in **Lyra (the Lyre)** and **Deneb** in **Cygnus (the Swan)**, form the well-known summer asterism. Though the asterism is called the Summer Triangle, it's actually visible until October. The three constellations will dominate the summer sky as they move higher up.

To the northeast is **Lacerta (the Lizard)**, an obscure kite-like constellation that lies between **Cassiopeia** and Cygnus. This region is part of the heart of the Milky Way and provides some beautiful binocular targets, including several open clusters. Small telescopes are needed to view a few of them, however. **NGC 7243** is an open star cluster near the kite-head of Lacerta. It is visible through binoculars and small telescopes.

This month is also the time for **noctilucent clouds**. These clouds form high in the atmosphere and occur above the northern horizon in the Northern Hemisphere beginning in mid-May. It's believed the clouds are due to meteoroid dust in the mesosphere, high in Earth's atmosphere. The clouds used to be restricted to more polar locations, but in recent years they have been spotted as far south as Los Angeles.

July

July's Events

Shorter nights might be a challenge, but there's no shortage of targets for you to take in this month.

The Milky Way is in the perfect position in this month. This region is chock-full of both open and globular clusters — some visible with nothing more than a pair of binoculars — as well as a stunning array of nebulae. Not to be left out is the Summer Triangle, an asterism involving the brightest stars of the constellations Cygnus, Aquila and Lyra (Deneb, Altair and Vega, respectively).

When it comes to the planets, there's a dance going on in the western sky after sunset.

Mars and Venus begin the month by edging closer and closer to one another, when finally, on July 13, they have their closest encounter. The pair will be roughly half a degree apart and low in the west just after sunset. They then begin to drift ever farther apart until Mars edges near the horizon by the end of the month.

There's also an event right here at home: on July 6 Earth is at aphelion, when it is the farthest from the Sun in its orbit. (Contrary to what some might believe, the Sun is not the farthest away from us in January.)

Calendar of Events

Day	Time (UTC)	Event
1	21:11	Last quarter of the Moon
4	19:59	Mercury 21.6°W of Sun (greatest western elongation)
5	14:48	Moon at apogee: 405,341 km (251,867 mi.)
6	02:59	Earth at aphelion: 1.0167 au (152,100,527 km)
8	04:38	Mercury 4.1°S of Moon
10	01:16	New Moon
12	09:10	Venus 3.5°S of Moon
12	10:10	Mars 4°S of Moon
13	13:17	Venus 0.5°N of Mars
17	10:11	First quarter of the Moon
21	10:30	Moon at perigee: 364,520 km (226,502 mi.)
21	21:21	Venus 1.1°N of Regulus
24	02:37	Full Moon
24	16:42	Saturn 3.9°N of Moon
26	01:17	Jupiter 4.3°N of Moon
29	14:09	Mars 0.6°N of Regulus
31	13:16	Last quarter of the Moon

The Moon This Month

SUN	MON	TUES	WED	THURS	FRI	SAT
				1 3rd Quarter	2	3
4	5	6	7	8	9	10 New Moon
11	12	13	14	15	16	17 1st Quarter
18	19	20	21	22	23	24 Full Moon
25	26	27	28	29	30	31 3rd Quarter

On July 12 there's a double planetary whammy, when both Mars and Venus will be roughly 6 degrees west of a thin crescent Moon that is sitting very low in the west after sunset. A low horizon with a clear sky will be essential to view this event. You could also try the night before, when the planets are about the same distance east of the Moon (for western observers, this is the closer approach).

On the night of July 25, Jupiter will be about 5 degrees north of the Moon, rising in the late-night sky. The planet remains visible in the southwestern sky just before sunrise.

The Moon is at apogee on July 5 and perigee on July 21. The new Moon occurs on July 10, and the full Moon on July 24.

The Summer Triangle asterism shines in the sky

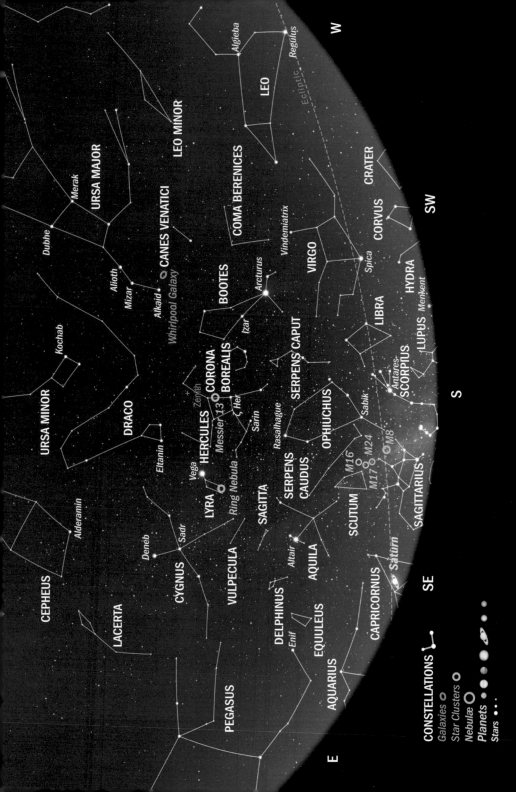

The Eagle Nebula (Messier 16)

Highlights in the Southern Sky

There is a lot to see in July. The constellation **Scorpius (the Scorpion)** has finally risen above the horizon. To the east of Scorpius is **Sagittarius (the Archer)**, a constellation rich with targets for the naked-eye, binocular or telescopic observer. It's quite easy to spot Sagittarius; just look for a teapot shape in the sky.

One of Sagittarius's most visible features, which can be viewed with the unaided eye particularly from a dark-sky location, is the **Sagittarius Star Cloud (Messier 24)**, found just north of the teapot's lid. This feature is located 10,000 light-years from Earth and is about 600 light-years wide. The cloud has an extremely dense collection of millions of stars.

Another object to look out for is the **Swan Nebula (Messier 17)**, also known as the Omega Nebula. This nebula is clearly visible in small telescopes and is also located in Sagittarius.

North of the Swan Nebula lies the **Eagle Nebula (Messier 16)**. This is the subject of the Hubble Space Telescope's famous photograph entitled "The Pillars of Creation." It is visible in large telescopes, and it is a favorite target of many astrophotographers.

The **Lagoon Nebula (Messier 8)**, also found in Sagittarius, is a wonderful emission nebula that you can see unaided under dark-sky conditions, though it will be faint. The nebula is located 5,200 light-years from Earth and has its own star cluster, which gives off incredible amounts of ultraviolet radiation and causes the nebula to shine.

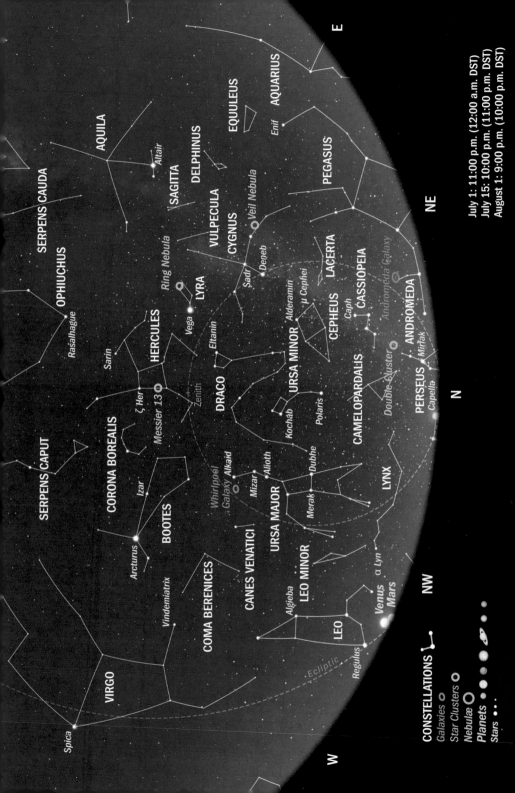

E

NE

N

NW

W

SERPENS CAUDA
OPHIUCHUS
AQUILA
Altair
SAGITTA
DELPHINUS
VULPECULA
EQUULEUS
AQUARIUS
Enf
Ring Nebula
CYGNUS
Veil Nebula
Deneb
Şadr
LYRA
Vega
PEGASUS
Eltanin
μ Cephei
LACERTA
CASSIOPEIA
Caph
ANDROMEDA
Andromeda Galaxy
Mirfak
PERSEUS
Capella
HERCULES
Sarin
ζ Her
Messier 13
Zenith
DRACO
URSA MINOR
Aldetamin
CEPHEUS
Double Cluster
Kochab
Polaris
CAMELOPARDALIS
Rasalhague
SERPENS CAPUT
CORONA BOREALIS
Izar
BOOTES
Arcturus
Whirlpool
Galaxy Alkaid
Mizar
Alioth
URSA MAJOR
Merak
Dubhe
LYNX
α Lyn
CANES VENATICI
LEO MINOR
COMA BERENICES
Vindemiatrix
Algieba
LEO
Venus
Mars
Regulus
VIRGO
Spica
Ecliptic

July 1: 11:00 p.m. (12:00 a.m. DST)
July 15: 10:00 p.m. (11:00 p.m. DST)
August 1: 9:00 p.m. (10:00 p.m. DST)

CONSTELLATIONS
Galaxies ○
Star Clusters ◯
Nebulæ ◯
Planets ● ● ●
Stars ● ● ·

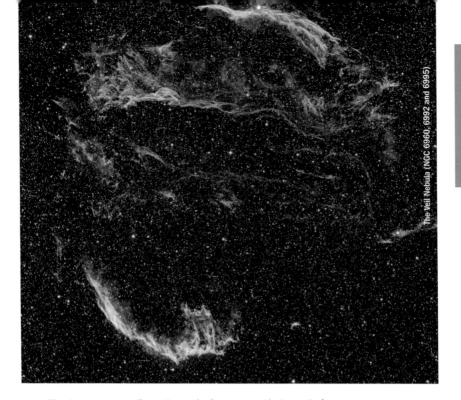

The Veil Nebula (NGC 6960, 6992 and 6995)

Highlights in the Northern Sky

The constellation **Camelopardalis (the Giraffe)** is in the northern sky, and **Cassiopeia** sits in the northeast. **Cepheus** is now high in the northeast, with the star **Mu Cephei**, or the Garnet Star – another spectacular target. The Garnet Star was given its name because of its deep reddish color. The star is variable, meaning that it fluctuates in brightness.

The bright stars **Deneb**, **Vega** and **Altair** form the **Summer Triangle**, which is now high in the sky and very visible. The constellation **Cygnus (the Swan)** contains numerous nebulae to look for, including the supernova remnant known as the **Veil Nebula (NGC 6960, 6992 and 6995)** – found southeast of **Sadr**, the center star in the "cross" of the constellation. The nebula is roughly 2,100 light-years away and is spread across the sky in an area that is equivalent to six Moons lined end-to-end. This nebula is a particular favorite of astrophotographers and is best seen with larger telescopes. It is actually made up of two parts, the Western Veil Nebula and the Eastern Veil Nebula.

The constellation **Pegasus (the Winged Horse)**, with its distinct square shape, is beginning to rise in the northeast, and **Delphinus (the Dolphin)**, sits high in the east.

The constellation **Ursa Major (the Great Bear)** is now beginning to sink low in the western sky, together with **Lynx (the Lynx)** and **Leo Minor (the Little Lion)**.

August

August's Events

This month we continue to enjoy the summer constellations and the warmer weather, as the nights are getting just a bit longer.

The biggest feature this month is undoubtedly the Perseid meteor shower. Many believe this to be the best shower of the year, as the weather tends to be clear and the outdoor temperatures enjoyable. The shower tends to be fairly consistent with bright fireballs. Though the shower began in July, the peak occurs on the night of August 12 to 13. On most nights, the shower is best at midnight and later, when the constellation Perseus is high in the sky,

Saturn is at opposition on August 2, and Jupiter is at opposition on August 19. This month they are at their brightest and largest – perfect for telescopic observation of satellites, rings, shadow transits and the Great Red Spot.

Calendar of Events

Day	Time (UTC)	Event
1	14:00	Mercury at superior conjunction
2	05:24	Saturn at opposition
2	07:35	Moon at apogee: 404,410 km (251,289 mi.)
8	13:50	New Moon
11	07:00	Venus 4.4°S of Moon
12–13		Perseid meteor shower peak
15	15:20	First quarter of the Moon
17	09:23	Moon at perigee: 369,124 km (229,363 mi.)
19	03:03	Mars 0.1°N of Mercury
19	23:05	Jupiter at opposition
20	22:19	Saturn 3.8°N of Moon
22	04:52	Jupiter 4.1°N of moon
22	12:02	Full Moon
30	02:22	Moon at apogee: 404,100 km (251,096 mi.)
30	07:13	Last quarter of the Moon

The Moon This Month

SUN	MON	TUES	WED	THURS	FRI	SAT
1	2	3	4	5	6	7
8 New Moon	9	10	11	12	13	14
15 1st Quarter	16	17	18	19	20	21
22 Full Moon	23	24	25	26	27	28
29	30 3rd Quarter	31				

The Perseids streak across the sky in mid-August

On the evenings of August 10 and 11, the crescent Moon will be about 7 degrees away from Venus in the western evening sky, first to the right and then to the left. Another after-sunset treat occurs on August 19, when Mars will be about 1 degree from Mercury very low in the west — a challenging near-horizon observation.

On August 20, Saturn will be about 4 degrees north of the waxing gibbous Moon in the night sky. And on the night of August 21, Jupiter will be about 5 degrees north of the nearly full Moon.

The Moon is at apogee on August 2, with the new Moon occurring on August 8. The first quarter is on August 15, followed by the full Moon a week later. The Moon will be at perigee on August 17 and then at apogee once again on August 30, which is also the date of the last quarter.

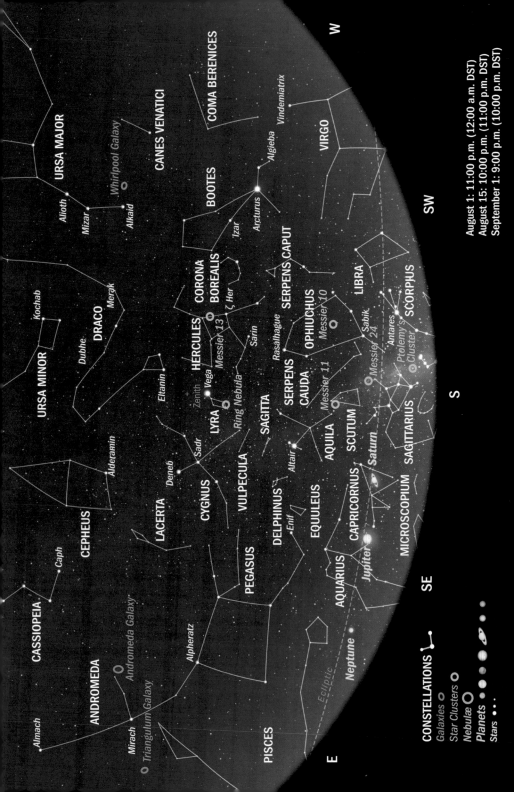

W

URSA MAJOR

Alioth
Mizar
Alkaid

COMA BERENICES

Whirlpool Galaxy

CANES VENATICI

BOOTES

Izar

Vindemiatrix

Algieba

Arcturus

VIRGO

SW

Kochab

Merak

Dubhe

DRACO

CORONA
BOREALIS

ζ Her

HERCULES

Messier 13

SERPENS CAPUT

Rasalhague

OPHIUCHUS

Messier 10

LIBRA

Sabik

Antares

Ptolemy's
Cluster

SCORPIUS

URSA MINOR

Etanin

Sarin

Zenith

Vega

LYRA

Sadr

Ring Nebula

SERPENS
CAUDA

Messier 11

Messier 24

S

CEPHEUS

Alderamin

Deneb

CYGNUS

VULPECULA

SAGITTA

AQUILA

SCUTUM

Altair

Saturn

SAGITTARIUS

August 1: 11:00 p.m. (12:00 a.m. DST)
August 15: 10:00 p.m. (11:00 p.m. DST)
September 1: 9:00 p.m. (10:00 p.m. DST)

LACERTA

DELPHINUS

Enif

EQUULEUS

MICROSCOPIUM

ANDROMEDA

Caph

CASSIOPEIA

Andromeda Galaxy

Alpheratz

PEGASUS

CAPRICORNUS

Jupiter

Almach

Mirach

Triangulum Galaxy

AQUARIUS

SE

PISCES

Neptune

Ecliptic

E

CONSTELLATIONS

Galaxies

Star Clusters

Nebulæ

Planets

Stars

The Sagittarius Star Cloud (Messier 24)

Highlights in the Southern Sky

The constellations **Sagittarius (the Archer)** and **Scorpius (the Scorpion)** still dominate the southern sky. Take a look at Sagittarius, and just off the tip of the teapot spout you'll find the center of our galaxy, the Milky Way. Below the spout lies **Ptolemy's Cluster (Messier 7)**, an open cluster of stars.

This is also a great time to enjoy the **Sagittarius Star Cloud (Messier 24)**. This cloud is a great binocular target, and it's easy to get lost among the stars.

The fantastic part about this time of year is that the Milky Way stretches right across the sky through the constellations **Cassiopeia, Cygnus (the Swan), Aquila (the Eagle)** and **Scutum (the Shield)** to Sagittarius. There are so many naked-eye targets, binocular targets and targets for telescopes of all sizes.

Pegasus (the Winged Horse) is now clear in the east, marked by "the Great Square," and **Delphinus (the Dolphin)** can be spotted between Pegasus and Aquila.

The constellation of Scutum can be found just above Sagittarius, and near Scutum's tip is the **Wild Duck Cluster (Messier 11)**. This open cluster of stars is located roughly 6,200 light-years from Earth and is one of the most densely populated open clusters known, containing approximately 2,900 stars.

The constellation **Ophiuchus (the Serpent Bearer)** rises above Scorpius, which puts it in a good place to observe the globular cluster **Messier 10**.

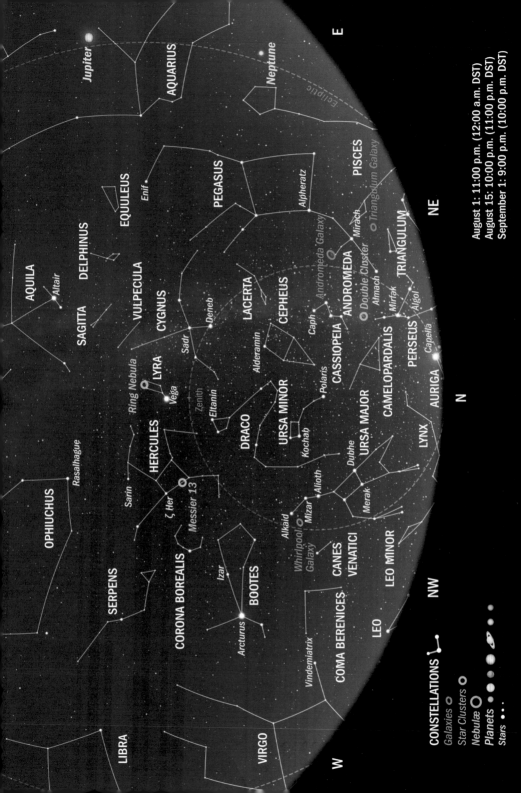

Highlights in the Northern Sky

The constellation **Ursa Major (the Great Bear)** is now low in the north-west, while **Camelopardalis (the Giraffe)** lies to the northeast.

Perseus, another constellation, is rising in the northeast with its brightest star **Mirfak**, which also means that the **Double Cluster (NGC 869 and NGC 884)** is rising along with it.

Cassiopeia shares the northwestern sky with its familiar "W" shape. But one of the best things to observe is the magnificent **Andromeda Galaxy (Messier 31)**. The Andromeda Galaxy is found easily by following the star **Shedar** in Cassiopeia to **Mirach** in the constellation **Andromeda** and then looking north. Andromeda is visible in moderate- to dark-sky locations as a faint fuzz. A pair of binoculars or a small telescope will reveal the nearest large galaxy to Earth in more detail, particularly the bright central bulge. Andromeda is 2.4 million light years from Earth, with a diameter of 200,000 light-years. This spiral galaxy is on a collision course with our Milky Way, though that event will likely take place 4.5 billion years from now.

Andromeda also has roughly 20 satellite galaxies; **Messier 32** and **Messier 110** are the best known.

The star **Arcturus** is now in the west, with the constellation **Hercules** above.

The Andromeda Galaxy (Messier 31)

September

September's Events

As the summer season comes to a close, we begin to bid farewell to the wealth of objects that drape across the center of our Milky Way. But there's still some time to soak it all in.

While Scorpius has now dipped below the horizon, Sagittarius is still visible, though low in the southwest. You also still have Jupiter and Saturn, both of which are prominent in the southern sky.

Let's not forget to enjoy the nearest spiral galaxy to us, the Andromeda Galaxy. This is beautiful when seen either with the naked eye at a dark-sky location, or alternatively, through binoculars or a small telescope.

On September 22, we welcome the fall season with the autumnal (fall) equinox.

Calendar of Events

Day	Time (UTC)	Event
5	14:32	Venus 1.6°N of Spica
7	00:52	New Moon
10	02:09	Venus 4.1°S of Moon
11	10:06	Moon at perigee: 368,461 km (228,951 mi.)
13	20:39	First quarter of the Moon
14	03:59	Mercury 26.8°E of Sun (greatest eastern elongation)
14	08:10	Neptune at opposition
17	02:37	Saturn 3.9°N of Moon
18	06:50	Jupiter 4.1°N of Moon
20	23:55	Full Moon
21	02:03	Mercury 1.4°S of Spica
22	19:21	Autumnal (fall) equinox
26	21:44	Moon at apogee: 404,640 km (251,432 mi.)
29	01:57	Last quarter of the Moon
30	14:00	Mercury 1.7°S of Spica

The Moon This Month

SUN	MON	TUES	WED	THURS	FRI	SAT
			1	2	3	4
5	6	7 New Moon	8	9	10	11
12	13 1st Quarter	14	15	16	17	18
19	20 Full Moon	21	22	23	24	25
26	27	28	29 3rd Quarter	30		

The beautiful Milky Way with its wealth of targets

The Moon and Venus have a close encounter on the evening of September 9, when the planet will be roughly 5 degrees southeast of the Moon.

Then, on the night of September 16, Saturn will be roughly 5 degrees north of the gibbous Moon. The next night, Jupiter will be about 6 degrees northeast of the Moon.

The new Moon occurs on September 7, with the perigee occurring on September 11. On September 20, we have the full Moon. Finally, we have the Moon at apogee on September 26, with the last quarter following three days later.

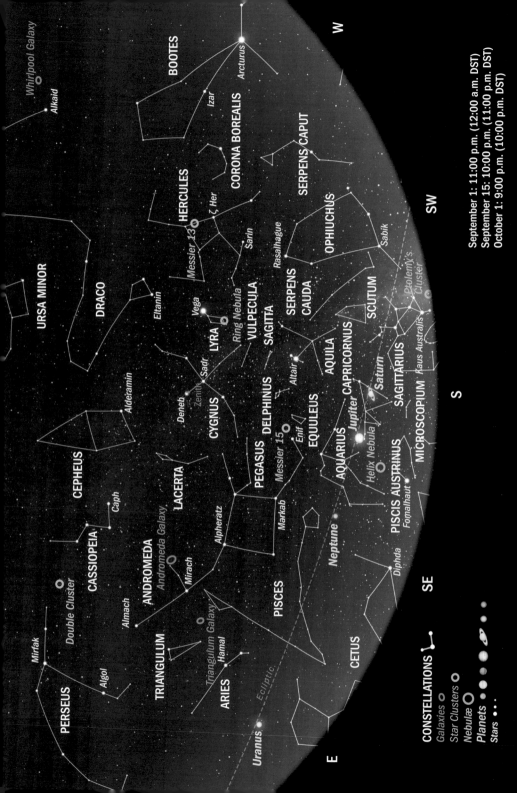

Highlights in the Southern Sky

Scorpius (the Scorpion) is now below the horizon, while **Sagittarius (the Archer)** is beginning to follow it in the southwestern sky.

The constellation **Aquarius (the Water Bearer)** is clearly visible in the southeast. An interesting target near Aquarius that can be seen through binoculars in dark-sky locations is the **Helix Nebula (NGC 7293)**, sometimes referred to as the Eye of God. This is a planetary nebula that formed after a star (similar to our own Sun) died and shed off its outer shell of gas. What's left behind is a small white-dwarf star. You can locate it by finding **Fomalhaut** in the constellation **Piscis Austrinus (the Southern Fish)**, which will be low on the southern horizon, then looking north 10 degrees.

The constellation **Capricornus (the Sea Goat)** lies in the south. You'll be able to find it quite easily this month, as **Jupiter** and **Saturn** will be situated on either side of it.

Equuleus (the Little Horse), a small constellation, can be found near **Delphinus (the Dolphin)**, north of Capricornus. To the east of Equuleus is **Messier 15**, another globular cluster that can be made out through binoculars under dark skies.

To the west of Delphinus is **Sagitta (the Arrow)**, another small constellation. Farther west of Sagitta is a wonderful asterism called the **Coathanger (Collinder 399, Brocchi's Cluster)**, which is best seen through binoculars or a small telescope. The asterism looks like an upside-down coat hanger and is difficult to miss.

The **Summer Triangle** now lies high in the south, and it is easy to see with the stars **Deneb**, **Vega** and **Altair** shining brightly. The constellation **Hercules** has made its way to the southwestern sky, with **Ophiuchus (the Serpent Bearer)** and **Serpens (the Serpent)** beginning their journey to the westward horizon. Serpens is the only constellation in two distinct parts: **Serpens Caput (the Head of the Serpent)** to the west of the Serpent Bearer, and **Serpens Cauda (the Tail of the Serpent)** to the east.

The Helix Nebula (NGC 7293)

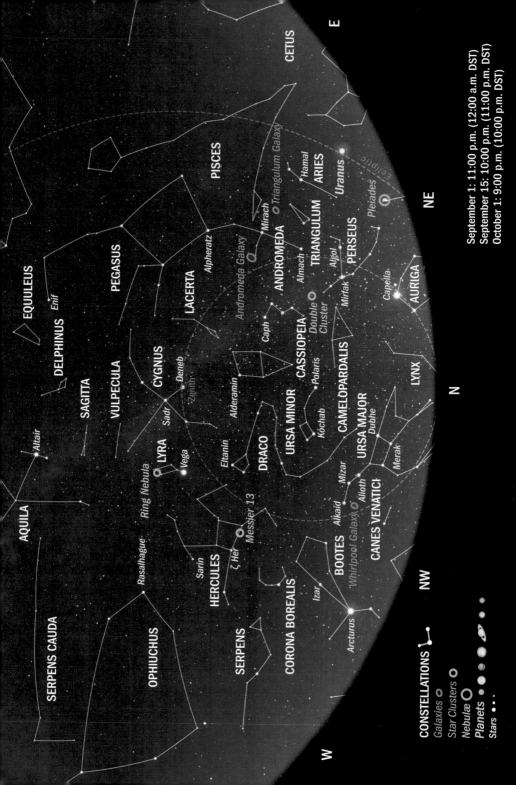

Highlights in the Northern Sky

The constellation **Boötes (the Herdsman)** is now low on the northwestern horizon, together with **Corona Borealis (the Northern Crown)**.

Hercules has moved to the northwestern sky and continues to provide a fantastic opportunity to observe the globular cluster **Messier 13**.

Meanwhile, **Ursa Major (the Great Bear)** is almost due north, with the **Big Dipper** asterism closer to the horizon. At the beginning of the month, the constellation **Auriga (the Charioteer)** is below the northeastern horizon, but by the end of the month it begins its northward journey, with its star **Capella** shining brightly near the horizon.

Perseus is in the northeast, and the **Double Cluster (NGC 869 and NGC 884)** sits between this constellation and **Cassiopeia**. The "house" that is **Cepheus** now appears upside-down in the north.

Mirfak, the brightest star in Perseus, is beautiful when seen through binoculars, particularly as there is a cluster of many bright stars in view, known as the **Alpha Persei Association (Melotte 20)**.

And of course, it's always worth observing the **Andromeda Galaxy (Messier 31)**, whether it's with the unaided eye in dark-sky conditions or through binoculars or a telescope.

The constellation Cassiopeia

September 2021

October

October's Events

We welcome the cooler, longer nights of October. As a reminder of the warmer months that have passed, the Summer Triangle is still visible, though now in the western sky. By the end of the month, Orion is beginning to peek out from the east, reminding us of the colder months ahead.

But the big story is the Orionid meteor shower, which will peak on the night of October 20 to 21. Though not spectacular, this shower can produce roughly 10 to 20 meteors an hour. The Southern Taurid meteor shower also peaks earlier in the month, although meteor activity for this shower can be seen into November.

Calendar of Events

Day	Time (UTC)	Event
6	11:05	New Moon
8	03:50	Mars in conjunction with Sun
8	17:28	Moon at perigee: 363,386 km (225,798 mi.)
9	16:12	Mercury at inferior conjunction
9	18:36	Venus 2.9°S of Moon
9–10		Southern Taurid meteor shower peak
13	03:25	First quarter of the Moon
14	07:12	Saturn 4.1°N of Moon
15	09:58	Jupiter 4.3°N of Moon
16	13:24	Venus 1.5°N of Antares
20	14:57	Full Moon
20–21		Orionid meteor shower peak
24	15:30	Moon at apogee: 405,615 km (252,037 mi.)
25	04:59	Mercury 18.4°W of Sun (greatest western elongation)
28	20:05	Last quarter of the Moon
29	21:59	Venus 47°E of Sun (greatest eastern elongation)

The Moon This Month

SUN	MON	TUES	WED	THURS	FRI	SAT
					1	2
3	4	5	6 New Moon	7	8	9
10	11	12	13 1st Quarter	14	15	16
17	18	19	20 Full Moon	21	22	23
24	25	26	27	28 3rd Quarter	29	30
31						

On October 9, Venus is roughly 2 degrees south of a crescent Moon, low in the southwest evening sky. On the night of October 13 to 14, Saturn will be roughly 6 degrees northeast of the waxing gibbous Moon. The following night we get a trio: the Moon will lie in between and slightly south of both Jupiter and Saturn.

On October 27, the waning gibbous Moon lies smack dab in the middle of Gemini (the Twins), visible in in the east late at night.

The new Moon occurs on October 6, with the Moon at perigee two days later. The full Moon occurs on October 20, and the Moon is at apogee on October 24.

A crescent Moon and Venus have a close pairing this month.

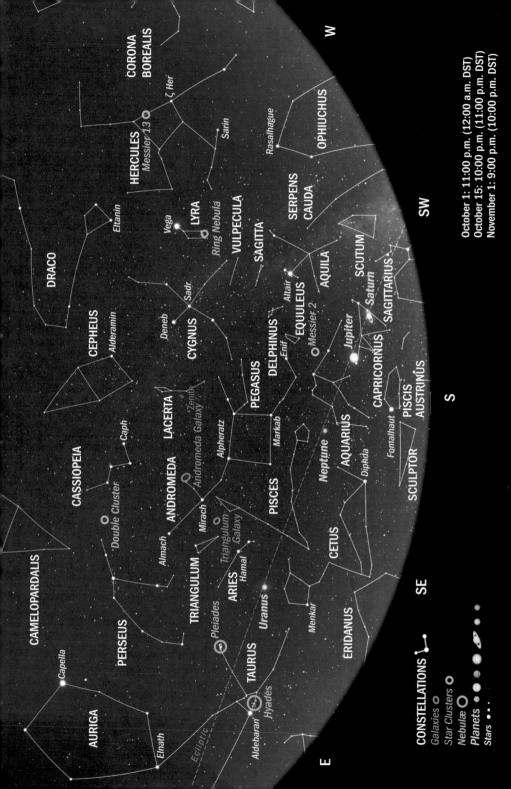

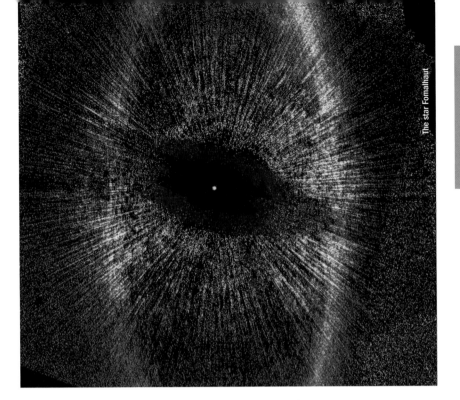

The star Fomalhaut

Highlights in the Southern Sky

The small constellation **Piscis Austrinus (the Southern Fish)** lies directly south, above the horizon, with its brightest star **Fomalhaut**. This star is located just 25 light-years from Earth, and it has a cloud of dust around it. In 2008 astronomers proposed that a planet lay within the cloud, based on data obtained by the Hubble Space Telescope; however, a new study in 2020 suggested instead that Hubble had witnessed two large asteroids.

The constellation **Aquarius (the Water Bearer)**, to the north, boasts a beautiful globular cluster, **Messier 2**. The cluster is within the constellation just below **Equuleus (the Little Horse)**, a small constellation north of Aquarius. The cluster is roughly 37,000 light-years from Earth, with a diameter that is approximately 150 light-years, making it one of the largest clusters visible in the sky.

Capricornus (the Sea Goat) lies to the west of Aquarius, with both **Jupiter** and **Saturn** still well-placed within the constellation.

Cetus (the Whale) is now clearly visible in the eastern sky along with **Aries (the Ram)** and **Pisces (the Fishes)**, above.

The mighty **Pegasus (the Winged Horse)** is now high in the southern sky, and to the west, the constellation **Aquila (the Eagle)** is beginning to set along with **Scutum (the Shield)**.

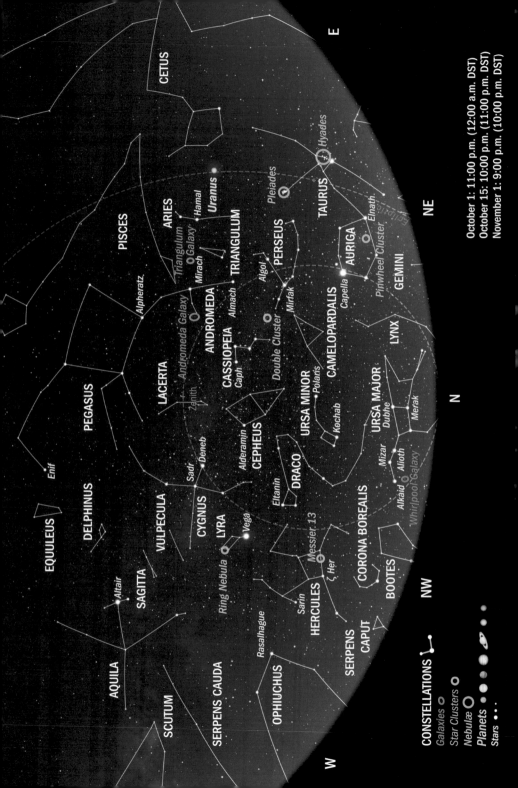

Highlights in the Northern Sky

Ursa Major (the Great Bear) is now directly north, with the **Big Dipper** asterism clearly visible, hugging the horizon.

Hercules is still well-placed in the west, while the constellations **Lyra (the Lyre)** and **Cygnus (the Swan)** are visible high in the west. It's still a good time to look at the globular cluster **Messier 13**, a great binocular target found in Hercules. And if you have a telescope, you should check out the **Ring Nebula (Messier 57)**, which lies in Lyra.

The star **Deneb**, at the tail of Cygnus, shines brightly high in the northwest and points toward **Cepheus**. At the center of Cygnus's cross, surrounding the star **Sadr**, is the **Sadr Region (IC 1318)**, a beautiful diffuse emission nebula.

Cassiopeia is well-placed high in the northeast together with **Perseus**, and **Auriga (the Charioteer)** is now visible in the east. The arrival of Auriga also provides a good opportunity to use binoculars to check out a few star clusters within it, including the **Pinwheel Cluster (Messier 36)**. Though not very bright, the cluster should be visible from dark-sky locations.

October 2021

The Sadr Region surrounding the star Sadr.

November

November's Events

While chillier weather may be settling across much of North America, the good news is that there's a lot more darkness to enjoy.

One of the best things about these chillier months is the return of Orion. The constellation has a lot to offer for both visual and telescopic observers, including the magnificent Orion Nebula.

Though the Southern Taurids peaked in mid-October, most observers won't see a huge difference in meteor rates between October and November. However, it should be noted that the Taurids in late October and early November tend to be brighter and often more visually spectacular. The Northern Taurids peak on November 12.

There's yet another shower that peaks a few days later from November 17 to 18, called the Leonids. This meteor shower has a ZHR of 15.

Finally, on November 19, North America is treated to a partial lunar eclipse. You can learn more about this on page 28.

Calendar of Events

Day	Time (UTC)	Event
4	21:15	New Moon
5	00:13	Uranus at opposition
5	22:23	Moon at perigee: 358,843 km (222,975 mi.)
7		Daylight Saving Time ends (most of U.S. and Canada)
8	05:21	Venus 1.1°S of Moon
10	14:27	Saturn 4.2°N of Moon
11	12:46	First quarter of the Moon
11	17:12	Jupiter 4.5°N of Moon
11–12		Northern Taurid meteor shower peak
17–18		Leonid meteor shower peak
19	08:58	Full Moon
19	09:02	Partial lunar eclipse
21	02:14	Moon at apogee: 406,279 km (252,450 mi.)
24	03:22	Pollux 2.8°N of Moon
27	12:28	Last quarter of the Moon
29	04:35	Mercury at superior conjunction

The Moon This Month

SUN	MON	TUES	WED	THURS	FRI	SAT
	1	2	3	4 New Moon	5	6
7	8	9	10	11 1st Quarter	12	13
14	15	16	17	18	19 Full Moon	20
21	22	23	24	25	26	27 3rd Quarter
28	29	30				

Venus and the Moon meet in the southwestern sky just after sunset on November 7. The planet will be roughly 5 degrees east of the Moon; the following night, the Moon will lie to the east of Venus.

Then, on the night of November 10, the waxing crescent Moon will be about 6 degrees southeast of Saturn. The next night, November 11, the first quarter Moon will be about 5 degrees south of Jupiter.

The first phase of the Moon this month is a new Moon on November 4. The next day, the Moon is at perigee, so seaside observers will experience higher ocean tides. The full Moon falls on November 19. Apogee occurs November 21, with the last quarter six days later.

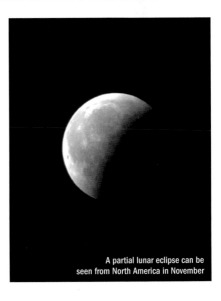

A partial lunar eclipse can be seen from North America in November

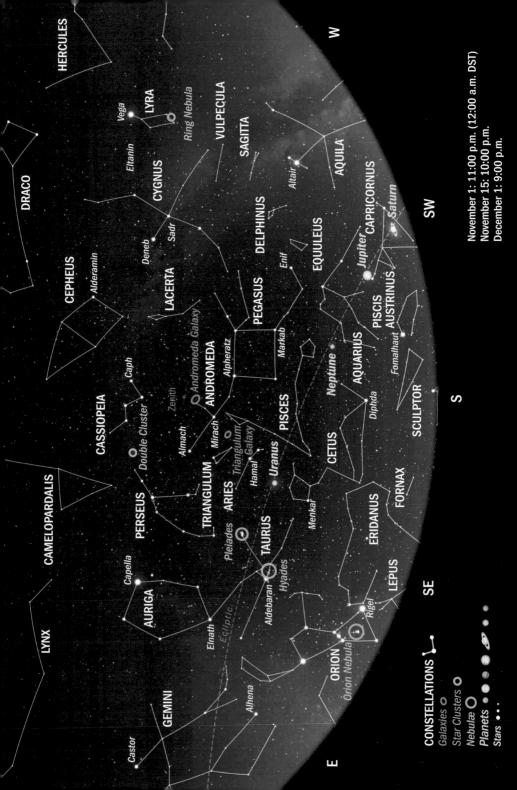

The Hyades (Melotte 25) and the Pleiades (Messier 45)

Highlights in the Southern Sky

The constellation **Cetus (the Whale)** takes center stage this month, lying directly south. **Aquarius (the Water Bearer)** begins to set in the west, and the square of **Pegasus (the Winged Horse)** is firmly high in the southwest and unmistakable.

But the highlight of the show is the reemergence of **Orion (the Hunter)** in the east. With the return of the familiar constellation, two amazing star clusters have also come back: the **Hyades (Melotte 25)** and the **Pleiades (Messier 45)**.

Both clusters are found in **Taurus (the Bull)**. The Hyades is a beautiful open star cluster in the head of the Bull, best viewed through binoculars, as are the Pleiades.

The brightest star in Taurus is **Aldebaran**, which lies approximately 65 light-years from Earth. It is about 44 times the radius of our Sun.

One of the longest constellations, **Eridanus (the River Eridanus)**, is now also above the horizon. Beginning near **Rigel** in Orion, the constellation represents a river in many cultures, past and present. In Ancient Egypt, it was the Nile or the Euphrates; for the Romans it was the Po; and in Ancient China it was the Huang He (also known as the Yellow River).

By the end of November, the constellation **Lepus (the Hare)** rises low on the southern horizon.

As for the planets, **Saturn** dips below the horizon in the late evening, and **Jupiter** is not far behind.

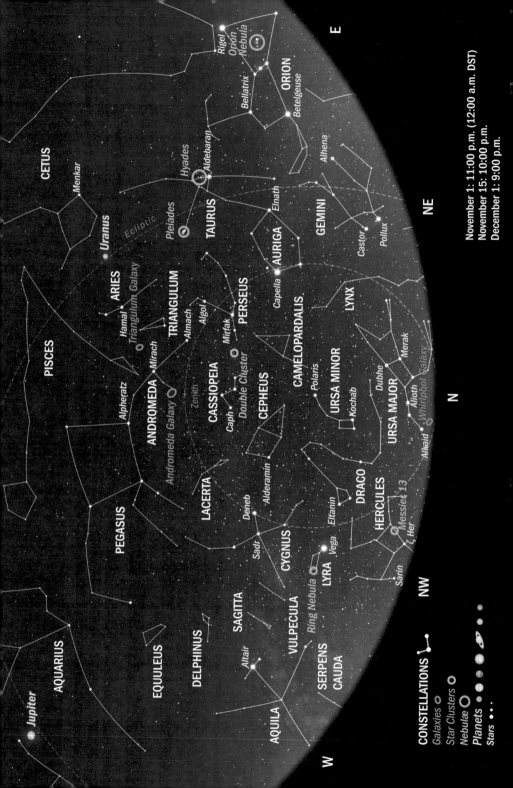

Highlights in the Northern Sky

In the west, the constellation **Hercules** is now below the horizon for the most part, and **Lyra (the Lyre)** is sinking along with **Cygnus (the Swan)**. The longer, cooler and clearer nights present a good opportunity to grab a pair of binoculars and take in the rich region of the Milky Way around Cygnus.

Draco (the Dragon) winds overhead to the north. **Ursa Major (the Great Bear)** starts its northward journey, and the first star in its tail — **Alkaid** — touches the northern horizon. **Ursa Minor (the Little Bear)**,

meanwhile, now hangs upside down by its tail.

Cassiopeia appears like an "M" rather than a "W" as it hangs high in the north, and the **Double Cluster (NGC 869 and NGC 884)** found near the constellation **Perseus** is directly overhead.

To the east, the **Gemini** twins are on their sides, with **Castor** and **Pollux** marking their heads, and the bright star **Capella**, belonging to **Auriga (the Charioteer)**, is clearly visible above.

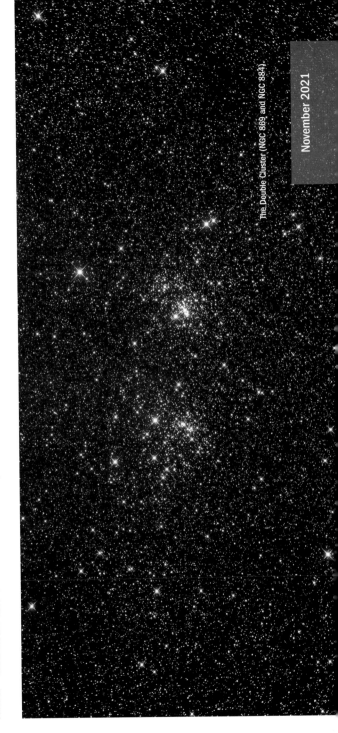

The Double Cluster (NGC 869 and NGC 884).

November 2021

December

December's Events

The big news in December is that Orion is back, but there are quite a few other events to enjoy during this wintry month.

On December 4, there's a total solar eclipse, though you'll have to go to Antarctica to see it in person. You can learn more about it on page 29.

The most active meteor shower of the year — the Geminids — is back to put on a show, peaking on the night of December 13 to 14. This meteor shower can be challenging to view only because of the weather, but it can produce upward of 150 meteors per hour under dark skies.

But that's not all we get in terms of meteor showers. The Ursids peak on the night of December 21 to 22. This shower is often forgotten as it is overshadowed by the very active Geminids, but it can produce 10 meteors per hour at its peak.

We ring in the last season of the year on December 21, with the winter solstice.

Calendar of Events

Day	Time (UTC)	Event
4		Total solar eclipse
4	07:43	New Moon
4	10:01	Moon at perigee: 356,794 km (221,702 mi.)
7	00:48	Venus 1.9°N of Moon
8	01:52	Saturn 4.2°N of Moon
9	06:07	Jupiter 4.6°N of Moon
11	01:36	First quarter of the Moon
13–14		Geminid meteor shower peak
18	02:16	Moon at apogee: 406,320 km (252,476 mi.)
19	04:36	Full Moon
21	09:20	Pollux 2.9°N of Moon
21	15:59	Winter solstice
21–22		Ursid meteor shower peak
27	02:24	Last quarter of the Moon
27	09:17	Mars 4.5°N of Antares
29	04:49	Venus 4.2°N of Mercury
31	20:13	Mars 0.9°N of Moon

The Moon This Month

SUN	MON	TUES	WED	THURS	FRI	SAT
			1	2	3	4 New Moon
5	6	7	8	9	10	11 1st Quarter
12	13	14	15	16	17	18
19 Full Moon	20	21	22	23	24	25
26 3rd Quarter	27	28	29	30	31	

Just after sunset on December 5, there is a wonderful alignment in the sky: From the south to the southwest are Jupiter, Saturn, Venus and a very thin crescent Moon (which might be difficult to see without the aid of binoculars and a low southwest horizon).

On December 6, we have a lovely pairing of Venus and the very thin crescent Moon, which will be roughly 10 percent illuminated in the western sky after sunset. Night by night, the Moon continues to march by Saturn and Jupiter in the evening sky.

The Moon is new on December 4, and also happens to be at perigee, which will give rise to large ocean tides in the following days for seaside observers. On December 11 is the first quarter, and a week later the Moon swings into apogee, with the full Moon close behind on December 19. The final last quarter of 2021 happens on December 27.

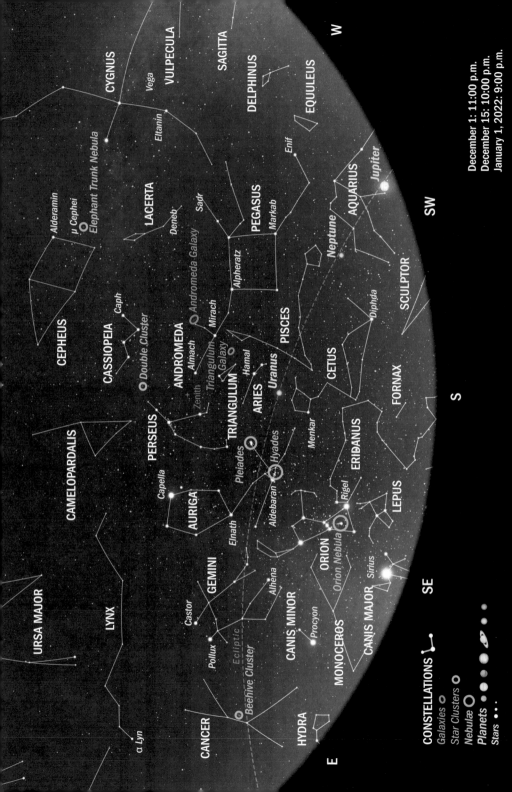

The Trapezium cluster, found inside the Orion Nebula (Messier 42)

Highlights in the Southern Sky

The winter southern sky can be just as impressive as the summer southern sky, and that's thanks mostly to **Orion (the Hunter)**.

This month, the Hunter is found high in the southeast. The **Orion Nebula (Messier 42)** is a favorite among many a stargazer. The nebula is visible to the unaided eye, even from light-polluted areas. Use a pair of binoculars, and you can make out the reddish color of this rich star-forming region. A telescope reveals much, much more, including the **Trapezium**, a cluster of four stars embedded in the nebula.

The constellation **Canis Major (the Great Dog)** follows behind Orion low in the southeast, with the brightest star in the night sky, **Sirius**, dominating. The constellation **Lepus (the Hare)** is now higher in the sky, and **Eridanus (the River Eridanus)** dominates the south.

Cetus (the Whale), another constellation, stretches from the south to southwest this month with **Aries (the Ram)** above. You might try to spot the aptly named constellation **Triangulum (the Triangle)**, just north of Aries. West of the Triangle lies the **Triangulum Galaxy (Messier 33)**, which is believed to be a satellite galaxy to the **Andromeda Galaxy (Messier 31)**. Under dark skies, some sharp-eyed observers can detect Messier 33 unaided, but binoculars should reveal it to everyone.

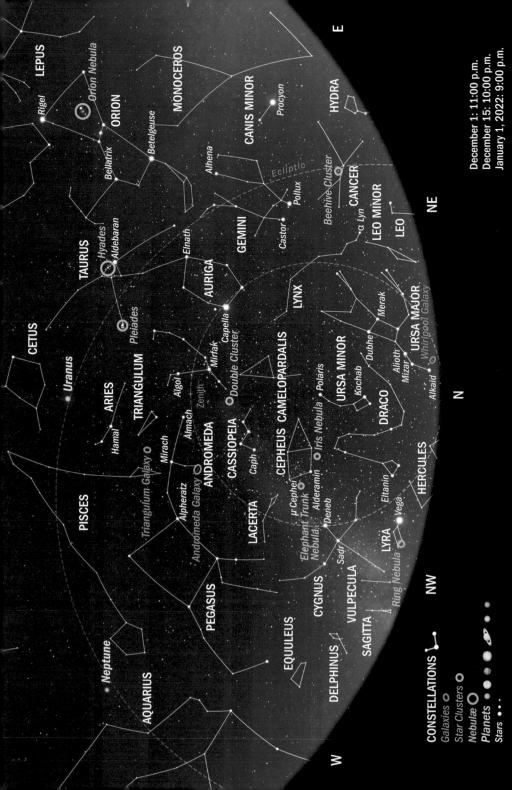

The Iris Nebula (NGC 7023)

Highlights in the Northern Sky

To the west, **Deneb** begins to slowly make its way toward the horizon, marking the disappearance of the constellation **Cygnus (the Swan)**. The constellation **Cepheus** begins its downward journey, trailing behind **Draco (the Dragon)**. Near Cepheus's **Mu Cephei** — which is still in a good position to observe — is the **Elephant Trunk Nebula (IC 1396A)**. This is a favorite target for many astrophotographers. Located 2,400 light-years from Earth, it is a rich star-forming region.

Also in Cepheus is another favorite: the **Iris Nebula (NGC 7023)**. This bright reflection nebula lies 1,300 light-years from Earth and is a stunning dusty region that resembles the iris flower, as its name suggests. **Cassiopeia** is sinking behind Cepheus, and the nearby **Andromeda Galaxy (Messier 31)** sits high in the northwest. Get to a dark-sky location or grab a pair of binoculars to enjoy our magnificent galactic neighbor.

The **Great Square of Pegasus** is high in the west and an unmistakable asterism. The constellation **Leo Minor (the Little Lion)** is low to the east, while **Lynx (the Lynx)** stretches above.

Glossary

annular solar eclipse: A special kind of partial solar eclipse during which the Moon does not completely cover the Sun's disk but leaves a ring of sunlight around the Moon.

aphelion: The point in a celestial object's orbit where it is farthest from the Sun.

apogee: When the Moon is at its most distant in its monthly orbit around Earth.

asterism: a group of stars within a constellation (or sometimes from several different constellations) that forms its own distinct pattern.

astronomical unit (au): A unit of distance that uses the average distance between Earth and the Sun as its base metric. One astronomical unit is approximately 150 million kilometers (93.2 million miles).

autumnal (fall) equinox: The date in September marking the end of summer and the beginning of autumn, when the Sun illuminates the Northern and Southern Hemispheres equally. The lengths of the day and night are equal.

binary star system: A star system in which two stars are gravitationally bound to each other and orbit a common center of mass.

conjunction: Technically speaking a conjunction is when two objects have the same right ascension, but amateur astronomers also use the term when there is a close approach of two or more celestial objects.

constellation: A group of stars that make an imaginary image in the night sky.

degree: A unit of measurement used to measure the distance between objects or the position of objects in astronomy. The entire sky spans 360 degrees, and up to about 180 degrees of sky is visible from any given point on Earth with an unobstructed horizon.

double star: A pair of stars that are spaced so closely together as to appear like a single star to an observer on Earth.

fireball: A very bright meteor, generally brighter than magnitude –4.0.

galaxy: A enormous system of gas, dust and billions of stars and their planetary systems, all held together by gravity.

globular cluster: Old star systems at the edges of spiral galaxies that can contain anywhere from thousands to millions of stars, packed in a close, roughly spherical form and held together by gravity.

greatest eastern elongation: When an inner planet (Mercury or Venus) is farthest from the Sun in the western evening sky.

greatest western elongation: When an inner planet (Mercury or Venus) is farthest from the Sun in the eastern morning sky.

light-year: A unit of distance that uses the distance light travels in one Earth year as its base metric. One light-year is about 9 trillion kilometers (6 trillion miles). Objects in outer space are so far apart that it takes a long time for their light to reach Earth, and so the farther an object is, the farther in the past that observers on Earth are seeing it. As an example, the Andromeda Galaxy (Messier 31) is 2.5 million light-years away, so when observers look at it in the sky, they are seeing it as it appeared 2.5 million years ago.

magnitude: The apparent brightness of an object in the sky. Magnitude is measured on a scale where the higher the number, the fainter the object appears. Objects with negative numbers are brighter than those with positive numbers.

meteor shower: A celestial event during which a number of meteors can be seen radiating from one point in the sky. Meteor showers occur when Earth, at a particular point in its orbit, crosses a stream of particles left over from a passing comet or asteroid.

meteor shower peak: The best time to view a meteor shower, when meteor activity will be at its highest.

nebula: A cloud of dust and gas visible in the night sky either as a bright patch or a dark shadow against other luminous matter. To learn more about the different types of nebulae, see page 36.

open cluster: A star system that contains anything from a dozen to hundreds of stars, in which the stars are spread out. Open clusters are found near the galactic plane, the plane on which most of a galaxy's mass lies.

opposition: When an outer planet (Mars, Jupiter, Saturn, Uranus or Neptune) is opposite the Sun in the sky in right ascension.

partial lunar eclipse: A lunar eclipse during which the Earth's shadow partially covers the Moon.

partial solar eclipse: A solar eclipse during which the Moon only partially covers the Sun.

perigee: When the Moon is at its closest in its monthly orbit around Earth.

perihelion: The point in a celestial object's orbit where it is closest to the Sun.

star: A luminous ball of plasma held together by its own gravity and fueled by the nuclear fusion of hydrogen into helium at its core. Stars come in a vast variety of sizes, luminosities and temperatures, typically ranging from dwarf sizes (that are as little as 10 percent the mass of the Sun) to hypergiants (that can be 100 or more times the mass of the Sun). To learn more about the different types of stars, see pages 10-11.

summer solstice: The point during the year that a particular hemisphere is most tilted toward the Sun. This takes place in June in the Northern Hemisphere.

supernova: A large explosion that occurs at the end of a star's life cycle during which its brightness increases and it throws off most of its mass.

total lunar eclipse: A lunar eclipse during which the Earth's shadow completely covers the Moon. Such an eclipse can last for hours.

total solar eclipse: A solar eclipse during which the Moon covers the entire face of the Sun, revealing the Sun's corona and prominences. Solar eclipses can last from a few seconds to about seven minutes maximum at a specific location.

variable star: A star whose apparent magnitude varies over time. The variability might be caused by physical changes to the star itself or by how it rotates or interacts with nearby objects.

vernal (spring) equinox: The date in March marking the end of winter and the beginning of spring, when the Sun illuminates the Northern and Southern Hemispheres equally. The lengths of the day and night are equal.

winter solstice: The point during the year that a particular hemisphere is most tilted away from the Sun. This takes place in December in the Northern Hemisphere.

Zenithal Hourly Rate (ZHR): The rate of meteors a shower would produce per hour under clear, dark skies and with the radiant at the zenith.

A breathtaking Milky Way scene over Binbrook, Ontario

Resources

American Astronomical Society — aas.org — An organization of professional astronomers seeking to enhance and share humanity's understanding of the Universe

American Meteor Society — www.amsmeteors.org — A non-profit scientific organization that informs, encourages and supports research in Meteor Astronomy

Eclipsewise.com — www.eclipsewise.com — Astronomer Fred Espenak's site dedicated to predictions and information on eclipses of the Sun and Moon

European Southern Observatory — www.eso.org — An intergovernmental research organization for ground-based astronomy that shares recent research, news and images

European Space Agency (ESA) — www.esa.int — The main site for the European Space Agency that shares news, research and launches

Hubblesite.org — hubblesite.org — NASA's main site for the Hubble Space Telescope, including news and images

NASA — www.nasa.gov — An update on all NASA missions, including astronomical events, news and launches

The Planetary Society — www.planetary.org — An international, non-profit organization that promotes the exploration of space through education, advocacy and research

The Royal Astronomical Society of Canada — www.rasc.ca — Home to Canada's national astronomical association

Spaceweather.com — spaceweather.com — An update on space weather, including solar flares, coronal mass ejections and noctilucent cloud forecasts

Space Weather Prediction Center — www.swpc.noaa.gov — The National Oceanic and Atmospheric Administration's site on forecasts for space weather, including potential geomagnetic storms

Solar and Heliospheric Observatory — soho.nascom.nasa.gov/home.html — A collaborative project between the ESA and NASA to study the Sun and its interactions with Earth from different perspectives

Solar Dynamics Observatory — sdo.gsfc.nasa.gov — Satellite observations of the Sun

Photo Credits

Alan Dyer: 26, 45, 47, 57, 59, 63, 85, 107

Alan Dyer/Stocktrek Images/Alamy Stock
Photo: 7

Debra Ceravolo: 5, 14 (right), 37 (top) 77, 91

Elliot Severn: 9

Fred Espenak: 27, 28 (bottom), 29 (top and
bottom)

Judy Anderson: 72

Kerry-Ann Lecky Hepburn: 113, 117

Malcolm Park: 15, 28 (top), 36, 48, 49, 51,
55, 73, 75, 79, 87, 95, 105

Michael Watson: 20–21

NASA: 14 (left), 34, 35, 53, 69, 99, 111

NASA/Bill Dunford: 18

NASA/Johns Hopkins University Applied
Physics Laboratory/Carnegie Institution of
Washington: 32

NASA/JPL: 33

NASA/SDO: 25

Nicole Mortillaro: 97, 103

Notanee Bourassa: 23, 31

Ron Brecher: 60, 65, 71, 81, 83, 93

Sean Walker: 24, 37 (bottom), 39 (top)

Stefanie Harron: 89, 101

Steven Castro/Shutterstock: 61

Yuichi Takasaka/Stocktrek Images/Alamy
Stock Photo: 42

Front cover: Alan Dyer/Stocktrek Images/
Alamy Stock Photo

You will also enjoy

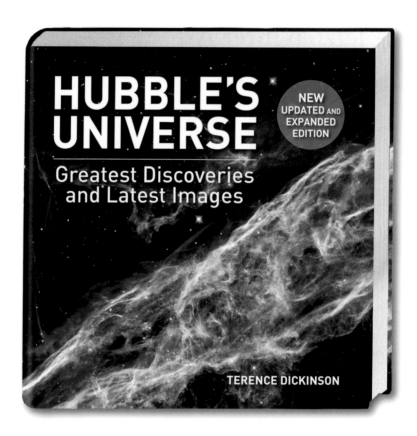

HUBBLE'S UNIVERSE 2nd edition

Greatest Discoveries and Latest Images

by Terence Dickinson, author of *NightWatch*

Updated and expanded with a spectacular gatefold
332 pages in glorious color

Wherever books are sold

FIREFLY BOOKS